THE PRACTICE OF BUSINESS STATISTICS

COMPANION CHAPTER 14
ONE-WAY ANALYSIS OF VARIANCE

David S. Moore
Purdue University

George P. McCabe
Purdue University

William M. Duckworth
Iowa State University

Stanley L. Sclove
University of Illinois

W. H. Freeman and Company
New York

Senior Acquisitions Editor:	**Patrick Farace**
Senior Developmental Editor:	**Terri Ward**
Associate Editor:	**Danielle Swearengin**
Media Editor:	**Brian Donnellan**
Marketing Manager:	**Jeffrey Rucker**
Head of Strategic Market Development:	**Clancy Marshall**
Project Editor:	**Mary Louise Byrd**
Cover and Text Design:	**Vicki Tomaselli**
Production Coordinator:	**Paul W. Rohloff**
Composition:	**Publication Services**
Manufacturing:	**RR Donnelley & Sons Company**

Cataloguing-in-Publication Data available from the Library of Congress

Library of Congress Control Number: 2002108463

Printed in the United States of America

Second printing

TO THE INSTRUCTOR

NOW *YOU* HAVE THE CHOICE!

This is **Companion Chapter 14** to *The Practice of Business Statistics (PBS).* Please note that this chapter, along with any other Companion Chapters, can be bundled with the *PBS* Core book, which contains Chapters 1–11.

CORE BOOK

Chapter 1	Examining Distributions
Chapter 2	Examining Relationships
Chapter 3	Producing Data
Chapter 4	Probability and Sampling Distributions
Chapter 5	Probability Theory
Chapter 6	Introduction to Inference
Chapter 7	Inference for Distributions
Chapter 8	Inference for Proportions
Chapter 9	Inference for Two-Way Tables
Chapter 10	Inference for Regression
Chapter 11	Multiple Regression

These other **Companion Chapters,** *in any combinations you wish,* are available for you to package with the *PBS* Core book.

TAKE YOUR PICK

Chapter 12	Statistics for Quality: Control and Capability
Chapter 13	Time Series Forecasting
Chapter 14	One-Way Analysis of Variance
Chapter 15	Two-Way Analysis of Variance
Chapter 16	Nonparametric Tests
Chapter 17	Logistic Regression
Chapter 18	Bootstrap Methods and Permutation Tests

ONE-WAY ANALYSIS OF VARIANCE

Prelude

How should we improve the forms?

A university requires its faculty to complete a form three times a year that describes the amount of time spent in various activities. For many years there were complaints that the form was difficult to fill out and that many of the questions were not clearly explained.

A new administration decided to revise the form. Several small groups of faculty members studied the old form and made recommendations for how it should be changed. From these groups, three possible versions of a new form were developed. To determine which of the new versions should be adopted, a randomized experiment was performed. From the list of faculty, 150 names were randomly selected. These were then randomly assigned to receive one of the three proposed forms. Thus, there were 50 faculty who received each form. The faculty were asked to review the proposed form and to rate it on a scale of 1 to 10.

The evaluations for the three proposed forms were compared using the analysis of variance methods described in this chapter. The results gave a clear indication that one of the forms was preferred to the other two. This form is now being used.

CHAPTER 14

One-Way Analysis of Variance

Introduction

Many of the most effective statistical studies are comparative, whether we compare the salaries of women and men in a sample of firms or the responses to two treatments in a clinical trial. We display comparisons with back-to-back stemplots or side-by-side boxplots, and we measure them with five-number summaries or with means and standard deviations. Now we ask: "Is the difference between groups statistically significant?"

When only two groups are compared, Chapter 7 provides the tools we need. Two-sample t procedures compare the means of two Normal populations, and we saw that these procedures, unlike comparisons of spread, are sufficiently robust to be widely useful. Now we will compare any number of means by techniques that generalize the two-sample t and share its robustness and usefulness.

14.1 One-Way Analysis of Variance

Which of four advertising offers mailed to sample households produces the highest sales in dollars? Which of ten brands of automobile tires wears longest? How long do cancer patients live under each of three therapies for their cancer? In each of these settings we wish to compare several treatments. In each case the data are subject to sampling variability—if we mailed the advertising offers to another set of households, we would get different data. We therefore pose the question for inference in terms of the *mean* response. We compare, for example, the mean tread lifetime of different brands of tires. In Chapter 7 we met procedures for comparing the means of two populations. We are now ready to extend those methods to problems involving more than two populations. The statistical methodology for comparing several means is called **analysis of variance,** or simply **ANOVA.**

ANOVA

one-way ANOVA

factor

We will consider two variations of the ANOVA idea. In **one-way ANOVA** we classify the populations of interest according to a single categorical explanatory variable that we call a **factor.** For example, to compare the tread lifetimes of ten specific brands of tires we use one-way ANOVA. This chapter presents the details for one-way ANOVA.

two-way ANOVA

In many practical situations we classify populations in more than one way. A mail-order firm might want to compare mailings that offer different discounts and also have different layouts. Will a lower price offered in a plain format draw more sales on the average than a higher price offered in a fancy brochure? Analyzing the effect of price and layout together requires **two-way ANOVA,** and adding yet more factors necessitates higher-way ANOVA techniques. Most of the new ideas in ANOVA with more than one factor already appear in two-way ANOVA, which we discuss in Chapter 15.

The ANOVA setting: comparing means

Do two population means differ? If we have random samples from the two populations, we compute a two-sample t statistic and its P-value to assess the statistical significance of the difference in the sample means. We compare several means in much the same way. Instead of a t statistic, ANOVA uses an F statistic and its P-value to evaluate the null hypothesis that all of several population means are equal.

In the sections that follow, we will examine the basic ideas of analysis of variance and the conditions under which we can use this tool. Although the details differ, many of the concepts are similar to those discussed in the two-sample case.

One-way analysis of variance is a statistical method for comparing several population means. We draw a simple random sample (SRS) from each population and use the data to test the null hypothesis that the population means are all equal.

EXAMPLE 14.1 **Comparing magazine covers**

A magazine publisher wants to compare three different styles of covers for a magazine that will be offered for sale at supermarket checkout lines. She assigns 60 stores at random to the three styles of covers and records the number of magazines that are sold in a one-week period.

EXAMPLE 14.2 **Who shops where?**

How do five bookstores in the same city differ in the demographics of their customers? A market researcher asks 50 customers of each store to respond to a questionnaire. One variable of interest is the customer's age.

These two examples have several similarities. In both there is a single quantitative response variable measured on several individuals; the individuals are stores in the first example and customers in the second. We expect the data to be approximately Normal and will consider a transformation if they are not. We wish to compare several populations, stores displaying three cover styles in the first example and customers of five bookstores in the second. There is also an important difference between the examples. Example 14.1 describes an **experiment** in which stores are randomly assigned to cover styles. Example 14.2 is an **observational study** in which customers are selected during a particular time period and not all agree to provide data. We will treat our samples of customers as random samples even though this is only approximately true. In both examples, we will use ANOVA to compare the mean responses. The same ANOVA methods apply to data from random samples and to data from randomized experiments. Do keep the data-production method in mind when interpreting the results, however. A strong case for causation is best made by a randomized experiment. We will often use the term *groups* for the populations to be compared in a one-way ANOVA.

experiment

observational study

To find out whether several populations all have the same mean, we will compare the means of samples drawn from each population.

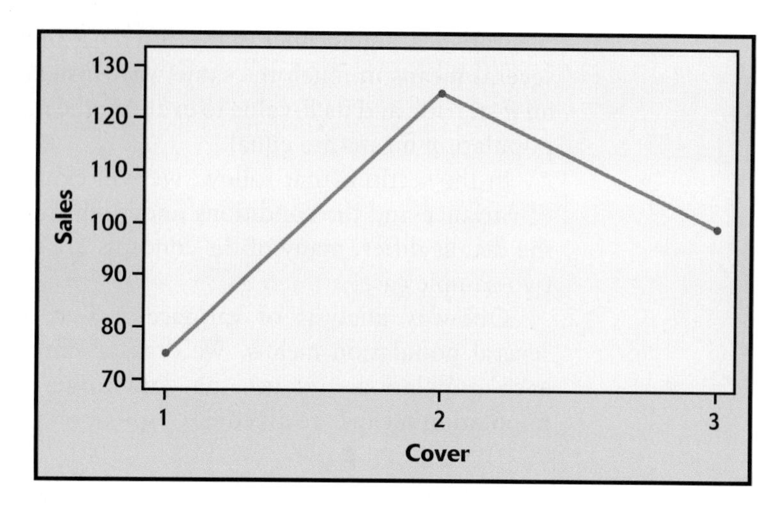

FIGURE 14.1 Mean sales of magazines for three cover designs.

Figure 14.1 displays the sample means for Example 14.1. Design 2 had the highest average sales. But is the observed difference among the designs just the result of chance variation? We do not expect sample means to be equal even if the population means are all identical. The purpose of ANOVA is to assess whether the observed differences among sample means are *statistically significant*. In other words, could a variation this large be plausibly due to chance, or is it good evidence for a difference among the population means? This question can't be answered from the sample means alone. Because the standard deviation of a sample mean $\bar{x}$ is the population standard deviation σ divided by $\sqrt{n}$, the answer depends upon both the variation within the groups of observations and the sizes of the samples.

Side-by-side boxplots help us see the within-group variation. Compare Figures 14.2(a) and 14.2(b). The sample medians are the same in both

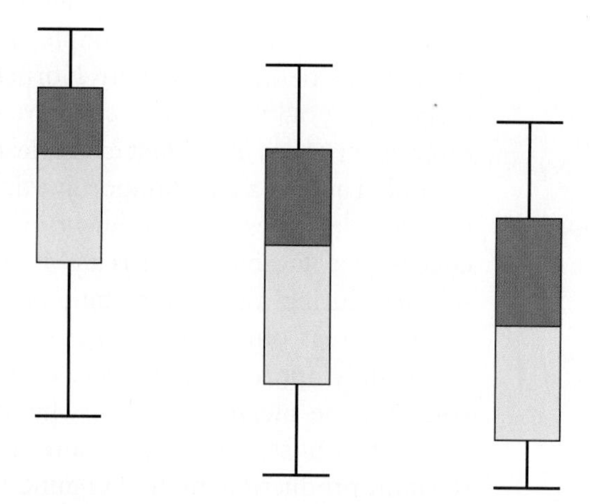

FIGURE 14.2(a) Side-by-side boxplots for three groups with large within-group variation. The differences among centers may be just chance variation.

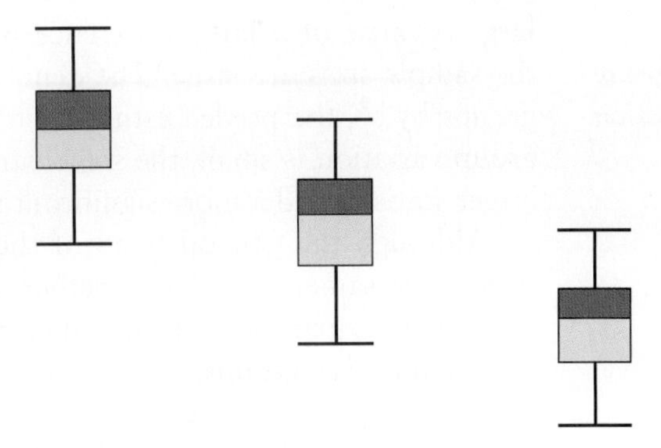

FIGURE 14.2(b) Side-by-side boxplots for three groups with the same centers as in Figure 14.2(a) but with small within-group variation. The differences among centers are more likely to be significant.

figures, but the large variation within the groups in Figure 14.2(a) suggests that the differences among the sample medians could be due simply to chance variation. The data in Figure 14.2(b) are much more convincing evidence that the populations differ. Even the boxplots omit essential information, however. To assess the observed differences, we must also know how large the samples are. Nonetheless, boxplots are a good preliminary display of ANOVA data. (ANOVA compares means, and boxplots display medians. If the distributions are nearly symmetric, these two measures of center will be close together.)

The two-sample t statistic

Two-sample t statistics compare the means of two populations. If the two populations are assumed to have equal but unknown standard deviations and the sample sizes are both equal to n, the t statistic is (page 464)

$$t = \frac{\overline{x} - \overline{y}}{s_p\sqrt{\frac{1}{n} + \frac{1}{n}}} = \frac{\sqrt{\frac{n}{2}}(\overline{x} - \overline{y})}{s_p}$$

The square of this t statistic is

$$t^2 = \frac{\frac{n}{2}(\overline{x} - \overline{y})^2}{s_p^2}$$

If we use ANOVA to compare two populations, the ANOVA F statistic is exactly equal to this t^2. We can therefore learn something about how ANOVA works by looking carefully at the statistic in this form.

between-group variation

The numerator in the t^2 statistic measures the variation **between** the groups in terms of the difference between their sample means $\overline{x}$ and $\overline{y}$. It includes a factor for the common sample size n. The numerator can be

within-group variation

large because of a large difference between the sample means or because the sample sizes are large. The denominator measures the variation **within** groups by s_p^2, the pooled estimator of the common variance. If the within-group variation is small, the same variation between the groups produces a larger statistic and a more significant result.

Although the general form of the F statistic is more complicated, the idea is the same. To assess whether several populations all have the same mean, we compare the variation *among* the means of several groups with the variation *within* groups. Because we are comparing variation, the method is called *analysis of variance*.

An overview of ANOVA

ANOVA tests the null hypothesis that the population means are *all equal*. The alternative is that they are not all equal. This alternative could be true because all of the means are different or simply because one of them differs from the rest. This is a more complex situation than comparing just two populations. If we reject the null hypothesis, we need to perform some further analysis to draw conclusions about which population means differ from which others.

The computations needed for ANOVA are more lengthy than those for the t test. For this reason we generally use software to perform the calculations. Automating the calculations frees us from the burden of arithmetic and allows us to concentrate on interpretation. The following example illustrates the practical use of ANOVA in analyzing data. Later we will explore the technical details.

EXAMPLE 14.3 Changing majors

In the computer science department of a large university, many students change their major after the first year. A detailed study[1] of the 256 students enrolled as first-year computer science majors in one year was undertaken to help understand this phenomenon. Students were classified on the basis of their status at the beginning of their second year, and several variables measured at the time of their entrance to the university were obtained. Here are summary statistics for the SAT mathematics scores:

Second-year major	n	$\bar{x}$	s
Computer science	103	619	86
Engineering and other sciences	31	629	67
Other	122	575	83

Figure 14.3 gives side-by-side boxplots of the SAT data. Compare Figure 14.3 with the plot of the mean scores in Figure 14.4. The means appear to be different, but there is a large amount of overlap in the three distributions. When we perform an ANOVA on these data, we ask a question about the group means. The null hypothesis is that the population mean SAT scores for the three groups are equal, and the alternative is that they are not all equal. The report on the study states that

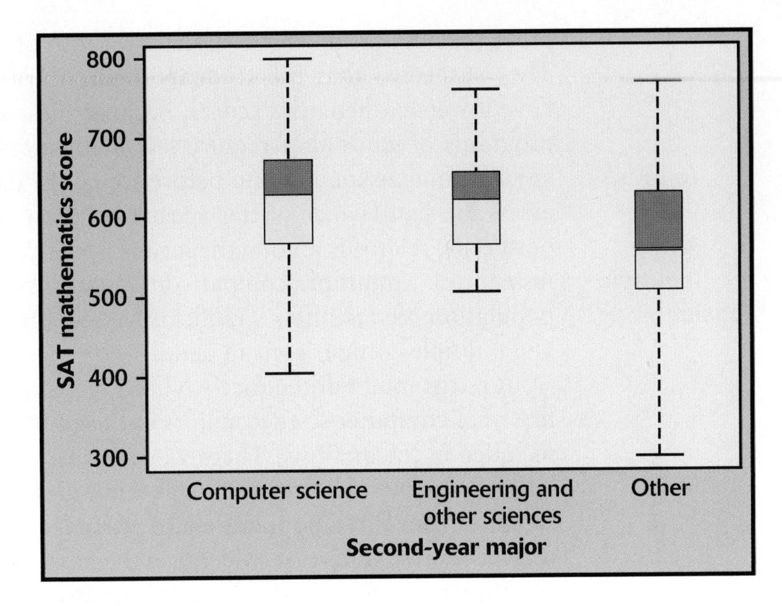

FIGURE 14.3 Side-by-side boxplots of SAT mathematics scores for the change-of-majors study.

the ANOVA F statistic is 10.35 with $P < 0.001$. There is very strong evidence that the three groups of students do not all have the same mean SAT mathematics scores.

Although we rejected the null hypothesis, we cannot conclude that all three population means are different, only that they are not all the same. Inspection of Figures 14.3 and 14.4 suggests that the means of the first two groups are nearly equal and that the third group differs from these two. We need to do some additional analysis to compare the three means. The researchers expected the mathematics scores of the engineering and

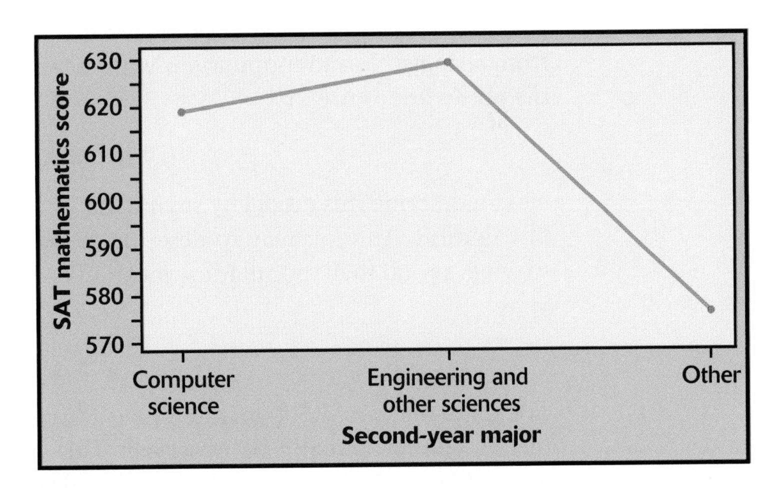

FIGURE 14.4 SAT mathematics score means for the change-of-majors study.

other sciences group to be similar to those of the computer science group. They suspected that the students in the third (other major) group would have lower mathematics scores, because these students chose to transfer to programs of study that require less mathematics. The researchers had this specific comparison in mind before seeing the data, so they used **contrasts** to assess the significance of this specific relation among the means. If we have no specific relations among the means in mind before looking at the data, we instead use a **multiple comparisons** procedure to determine which pairs of population means differ significantly. Section 14.2 explores both contrasts and multiple comparisons in detail.

contrasts

multiple comparisons

A purist might argue that ANOVA is inappropriate in Example 14.3. *All* first-year computer science majors in this university for the year studied were included in the analysis. There was no random sampling from larger populations. On the other hand, we can think of these three groups of students as samples from three populations of students in similar circumstances. They may be representative of students in the next few years at the same university or other universities with similar programs. Judgments such as this are very important and must be made, not by statisticians, but by people who are knowledgeable about the subject of the study.

The ANOVA model

When analyzing data, we think in terms of an overall pattern and deviations from it. In shorthand form,

$$\text{DATA} = \text{FIT} + \text{RESIDUAL}$$

In the regression model of Chapter 10, the FIT was the population regression line, and the RESIDUAL represented the deviations of the data from this line. We now apply this framework to describe the statistical models used in ANOVA. These models provide a convenient way to summarize the conditions that are the foundation for our analysis. They also give us the necessary notation to describe the calculations needed.

First, recall the statistical model for a random sample of observations from a single Normal population with mean μ and standard deviation σ. If the observations are

$$x_1, x_2, \ldots, x_n$$

we can describe this model by saying that the x_j are an SRS from the $N(\mu, \sigma)$ distribution. Another way to describe the same model is to think of the x's varying about their population mean. To do this, write each observation x_j as

$$x_j = \mu + \epsilon_j$$

The ϵ_j are then an SRS from the $N(0, \sigma)$ distribution. Because μ is unknown, the ϵ's cannot actually be observed. This form more closely corresponds to our

$$\text{DATA} = \text{FIT} + \text{RESIDUAL}$$

way of thinking. The FIT part of the model is represented by μ. It is the systematic part of the model, like the line in a regression. The RESIDUAL part is represented by ϵ_j. It represents the deviations of the data from the fit and is due to random, or chance, variation.

There are two unknown parameters in this statistical model: μ and σ. We estimate μ by $\overline{x}$, the sample mean, and σ by s, the sample standard deviation. The differences $e_j = x_j - \overline{x}$ are the sample residuals. They correspond to the ϵ_j in the statistical model.

The model for one-way ANOVA is very similar. We take random samples from each of I different populations. The sample size is n_i for the ith population. Let x_{ij} represent the jth observation from the ith population. The I population means are the FIT part of the model and are represented by μ_i. The random variation, or RESIDUAL, part of the model is represented by the deviations ϵ_{ij} of the observations from the means.

THE ONE-WAY ANOVA MODEL

The data for one-way ANOVA are SRSs from each of I populations. The sample from the ith population has n_i observations, $x_{i1}, x_{i2}, \ldots, x_{in_i}$. The **one-way ANOVA model** is

$$x_{ij} = \mu_i + \epsilon_{ij}$$

for $i = 1, \ldots, I$ and $j = 1, \ldots, n_i$. The ϵ_{ij} are assumed to be from an $N(0, \sigma)$ distribution. The **parameters of the model** are the I population means $\mu_1, \mu_2, \ldots, \mu_I$ and the common standard deviation σ.

The sample sizes n_i may differ, but the standard deviation σ is assumed to be the same in all of the populations. Figure 14.5 pictures this model for $I = 3$. The three population means μ_i are different, but the shapes of the three Normal distributions are the same, reflecting the condition that all three populations have the same standard deviation.

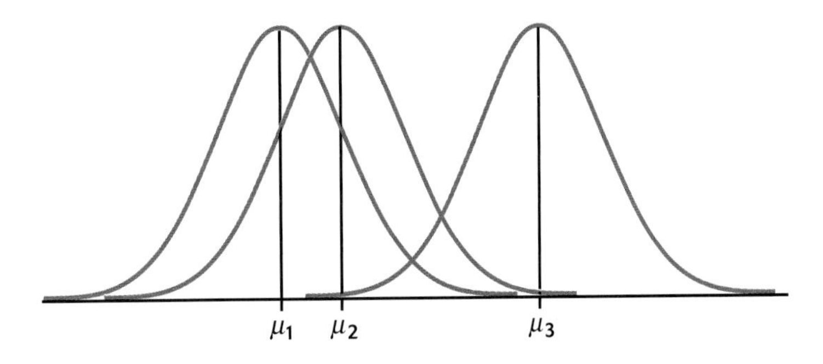

FIGURE 14.5 Model for one-way ANOVA with three groups. The three populations have Normal distributions with the same standard deviation.

EXAMPLE 14.4 **How much do students spend on textbooks?**

A survey of college students attempted to determine how much money they spend per year on textbooks and to compare students by year of study. From lists of students provided by the registrar, SRSs of size 50 were chosen from each of the four classes (freshmen, sophomores, juniors, and seniors). The students selected were asked how much they had spent on textbooks during the current semester.

There are $I = 4$ populations. The population means μ_1, μ_2, μ_3, and μ_4 are the average amounts spent on textbooks by *all* freshmen, sophomores, juniors, and seniors at this college for this semester. The sample sizes n_i are 50, 50, 50, and 50.

Suppose the first freshman sampled is Eve Brogden. The observation x_{11} is the amount spent by Eve. The data for the other freshmen sampled are denoted by x_{12}, x_{13}, and so on. Similarly, the data for the other groups have a first subscript indicating the group and a second subscript indicating the student in the group.

According to our model, Eve's spending is $x_{11} = \mu_1 + \epsilon_{11}$, where μ_1 is the average for *all* of the students in the freshman class and ϵ_{11} is the chance variation due to Eve's specific needs. We are assuming that the ϵ_{ij} are independent and Normally distributed (at least approximately) with mean 0 and standard deviation σ.

APPLY YOUR KNOWLEDGE

14.1 **Magazine covers.** Example 14.1 (page 14-5) describes a study designed to compare sales based on different magazine covers. Write out the ANOVA model for this study. Be sure to give specific values for I and the n_i. List all of the parameters of the model.

14.2 **Ages of customers at different stores.** In Example 14.2 (page 14-5) the ages of customers at different bookstores are compared. Write out the ANOVA model for this study. Be sure to give specific values for I and the n_i. List all of the parameters of the model.

Estimates of population parameters

The unknown parameters in the statistical model for ANOVA are the I population means μ_i and the common population standard deviation σ. To estimate μ_i we use the sample mean for the ith group:

$$\bar{x}_i = \frac{1}{n_i}\sum_{j=1}^{n_i} x_{ij}$$

residuals

The **residuals** $e_{ij} = x_{ij} - \bar{x}_i$ reflect the variation about the sample means that we see in the data.

The ANOVA model states that the population standard deviations are all equal. The ANOVA test of equality of means requires this condition. If we have unequal standard deviations, we generally try to transform the data so that they are approximately equal. We might, for example, work with $\sqrt{x_{ij}}$ or $\log x_{ij}$. Fortunately, we can often find a transformation that *both* makes the group standard deviations more nearly equal and also makes the distributions of observations in each group more nearly Normal. If the standard deviations are markedly different and cannot be made similar by a transformation, inference requires different methods.

Unfortunately, formal tests for the equality of standard deviations in several groups share the lack of robustness against non-Normality that we noted in Chapter 7 for the case of two groups. Because ANOVA procedures are not extremely sensitive to unequal standard deviations, we do *not* recommend a formal test of equality of standard deviations as a preliminary to the ANOVA. Instead, we will use the following rule of thumb.

RULE FOR EXAMINING STANDARD DEVIATIONS IN ANOVA

If the largest sample standard deviation is less than twice the smallest sample standard deviation, we can use methods based on the condition that the population standard deviations are equal and our results will still be approximately correct.[2]

When we assume that the population standard deviations are equal, each sample standard deviation is an estimate of σ. To combine these into a single estimate, we use a generalization of the pooling method introduced in Chapter 7.

POOLED ESTIMATOR OF σ

Suppose we have sample variances $s_1^2, s_2^2, \ldots, s_I^2$ from I independent SRSs of sizes $n_1, n_2, \ldots, n_I$ from populations with common variance σ^2. The **pooled sample variance**

$$s_p^2 = \frac{(n_1 - 1)s_1^2 + (n_2 - 1)s_2^2 + \cdots + (n_I - 1)s_I^2}{(n_1 - 1) + (n_2 - 1) + \cdots + (n_I - 1)}$$

is an unbiased estimator of σ^2. The **pooled standard error**

$$s_p = \sqrt{s_p^2}$$

is the estimate of σ.

Pooling gives more weight to groups with larger sample sizes. If the sample sizes are equal, s_p^2 is just the average of the I sample variances. Note that s_p is *not* the average of the I sample standard deviations.

EXAMPLE 14.5 **Estimates of parameters for the change-of-majors study**

In the change-of-majors study of Example 14.3 there are $I = 3$ groups and the sample sizes are $n_1 = 103$, $n_2 = 31$, and $n_3 = 122$. The sample means for the SAT mathematics scores are $\bar{x}_1 = 619$, $\bar{x}_2 = 629$, and $\bar{x}_3 = 575$. These statistics estimate the three unknown population means.

The sample standard deviations are $s_1 = 86$, $s_2 = 67$, and $s_3 = 83$. Because the largest standard deviation (86) is less than twice the smallest ($2 \times 67 = 134$), our rule of thumb allows us to act as if the population standard deviations are equal.

The pooled variance estimate is

$$s_p^2 = \frac{(n_1 - 1)s_1^2 + (n_2 - 1)s_2^2 + (n_3 - 1)s_3^2}{(n_1 - 1) + (n_2 - 1) + (n_3 - 1)}$$

$$= \frac{(102)(86)^2 + (30)(67)^2 + (121)(83)^2}{102 + 30 + 121}$$

$$= \frac{1{,}722{,}631}{253} = 6809$$

The pooled standard error is

$$s_p = \sqrt{6809} = 82.5$$

This is our estimate of the common standard deviation σ of the SAT mathematics scores in the three populations of students.

14.3 **Magazine covers.** Example 14.1 (page 14-5) describes a study designed to compare sales based on different magazine covers, and in Exercise 14.1 you described the ANOVA model for this study. The three covers are designated A, B, and C. The following table summarizes the sales data.

Design	Mean	Standard deviation	Sample size
1	75	80	20
2	125	100	20
3	100	120	20

(a) Is it reasonable to pool the standard deviations for these data?

(b) For each parameter in your model from Exercise 14.1, give the estimate.

14.4 **Ages of customers at different stores.** In Example 14.2 (page 14-5) the ages of customers at different bookstores are compared, and you described the ANOVA model for this study in Exercise 14.2 (page 14-12). Here is a summary of the ages of the customers:

Store	Mean	Standard deviation	Sample size
A	34	8	50
B	42	12	50
C	20	7	50
D	22	8	50
E	46	14	50

(a) Is it reasonable to pool the standard deviations for these data?

(b) For each parameter in your model from Exercise 14.4, give the estimate.

Testing hypotheses in one-way ANOVA

Comparison of several means is accomplished by using an F statistic to compare the variation among groups with the variation within groups. We now show how the F statistic expresses this comparison. Calculations are

ANOVA table organized in an **ANOVA table,** which contains numerical measures of the variation among groups and within groups.

First we must specify our hypotheses for one-way ANOVA. As usual, I represents the number of populations to be compared.

HYPOTHESES FOR ONE-WAY ANOVA

The **null and alternative hypotheses** for one-way ANOVA are

$$H_0: \mu_1 = \mu_2 = \cdots = \mu_I$$
$$H_a: \text{not all of the } \mu_i \text{ are equal}$$

The example and discussion that follow illustrate how to do a one-way ANOVA. The calculations are generally performed using statistical software on a computer, so we focus on interpretation of the output.

CASE 14.1

A NEW EDUCATIONAL PRODUCT

Your company markets educational materials aimed at parents of young children. You are planning a new product that is designed to improve children's reading comprehension. Your product is based on new ideas from educational research, and you would like to claim that children will have better reading comprehension skills with the new methods than with the traditional approach. Your marketing material will include the results of a study conducted to compare two versions of the new approach with the traditional method.[3] The standard method is called Basal, and the two variations of the new method are called DRTA and Strat (for strategies).

Education researchers randomly divided 66 children into three groups of 22. Each group was taught by one of the three methods. As is common in such studies, the children were given a "pretest" that measured their reading comprehension before receiving any instruction. One purpose of the pretest was to see if the three groups of children were similar in their comprehension skills at the start of the study. One of the pretest variables was an "intruded sentences" measure, which measures one type of reading comprehension skill. The data for the 22 subjects in each group are given in Table 14.1. The three groups are named Basal, DRTA, and Strat.

We want to test the null hypothesis that three groups represent three populations that all have the same mean score on the pretest. First, look at the data. Side-by-side boxplots do not give a good picture of these data. This is a consequence of the relatively small number of observations in each group (22) combined with the fact that the variable has only a few possible values (the values are all integers between 4 and 17). Side-by-side stemplots give a better picture. After inspecting the stemplots, decide whether the ANOVA model should be used to analyze the data.

TABLE 14.1 Pretest reading scores

Group	Subject	Score	Group	Subject	Score	Group	Subject	Score
Basal	1	4	DRTA	23	7	Strat	45	11
Basal	2	6	DRTA	24	7	Strat	46	7
Basal	3	9	DRTA	25	12	Strat	47	4
Basal	4	12	DRTA	26	10	Strat	48	7
Basal	5	16	DRTA	27	16	Strat	49	7
Basal	6	15	DRTA	28	15	Strat	50	6
Basal	7	14	DRTA	29	9	Strat	51	11
Basal	8	12	DRTA	30	8	Strat	52	14
Basal	9	12	DRTA	31	13	Strat	53	13
Basal	10	8	DRTA	32	12	Strat	54	9
Basal	11	13	DRTA	33	7	Strat	55	12
Basal	12	9	DRTA	34	6	Strat	56	13
Basal	13	12	DRTA	35	8	Strat	57	4
Basal	14	12	DRTA	36	9	Strat	58	13
Basal	15	12	DRTA	37	9	Strat	59	6
Basal	16	10	DRTA	38	8	Strat	60	12
Basal	17	8	DRTA	39	9	Strat	61	6
Basal	18	12	DRTA	40	13	Strat	62	11
Basal	19	11	DRTA	41	10	Strat	63	14
Basal	20	8	DRTA	42	8	Strat	64	8
Basal	21	7	DRTA	43	8	Strat	65	5
Basal	22	9	DRTA	44	10	Strat	66	8

APPLY YOUR KNOWLEDGE

14.5 **Stemplots.** Display the data in Table 14.1 for the three reading groups using three stemplots. Make the stems the same for the groups and put them side by side so that the data can be compared. Are any systematic differences among the groups apparent? Do any groups show strong skewness or clear outliers?

CASE 14.1

EXAMPLE 14.6

Verifying the conditions for ANOVA

If ANOVA is to be trusted, three conditions must hold.

SRSs. Can we regard the three samples as SRSs from three populations? An ideal study would start with an SRS of all children of the same age and randomly assign children to the three teaching methods. This isn't practical. The researchers randomly assigned children from one school district. They argue that for comparing reading methods we can act as if these children were randomly chosen. Critics may disagree.

Normality. Is the response variable Normally distributed in each group? Figure 14.6 displays Normal quantile plots for the three groups. The data do look reasonably Normal. There are only a small number of possible values for the score (the values are all integers between 4 and 17), but this should not cause difficulties in using ANOVA.

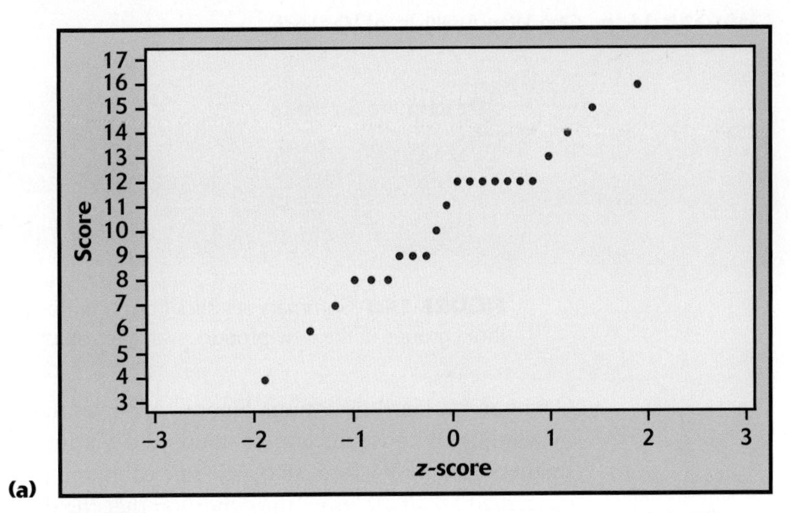

(a)

FIGURE 14.6 Normal quantile plots of the pretest scores in the (a) Basal, (b) DRTA, and (c) Strat groups in the new-product evaluation study.

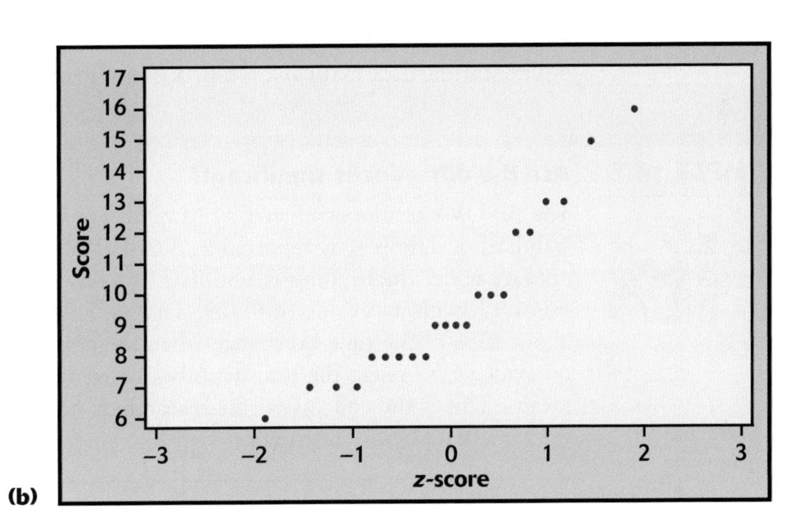

(b)

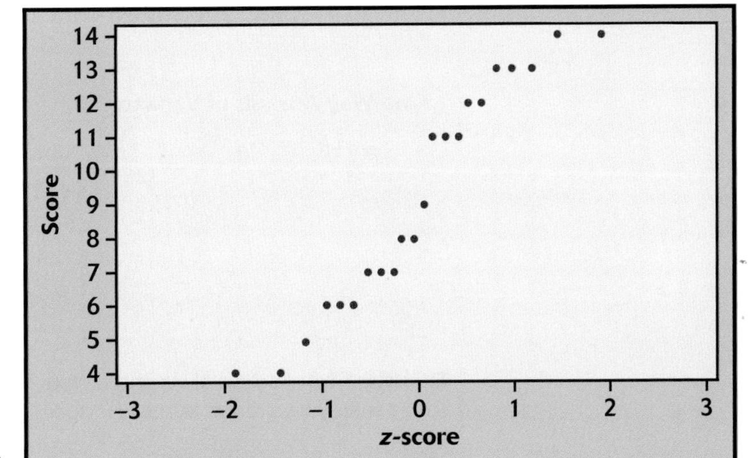

(c)

Descriptive Statistics

Variable	Group	N	Mean	Median	StDev
Score	Basal	22	10.500	11.500	2.972
	DRTA	22	9.727	9.000	2.694
	Strat	22	9.136	8.500	3.342

FIGURE 14.7 Summary statistics from Minitab for the pretest scores in the three groups in the new-product evaluation study.

Common standard deviation. Figure 14.7 gives the summary statistics generated by Minitab. Because the largest standard deviation (3.342) is less than twice the smallest ($2 \times 2.694 = 5.388$), our rule of thumb tells us that we need not be concerned about violating the condition that the three populations have the same standard deviation.

Because the data look reasonably Normal and meet the condition of equal standard deviations, we proceed with the analysis of variance.

EXAMPLE 14.7 **Are the differences significant?**

CASE14.1

The ANOVA results produced by Minitab are shown in Figure 14.8. The pooled standard deviation s_p is reported as 3.014. The calculated value of the F statistic appears under the heading F, and its P-value is under the heading P. The value of F is 1.13, with a P-value of 0.329. That is, an F of 1.13 or larger would occur about 33% of the time by chance when the population means are equal. We have no evidence to reject the null hypothesis that the three populations have equal means. This is the conclusion the researchers hoped for—the three groups are on the same level before instruction begins.

Examples 14.6 and 14.7 illustrate the basics of ANOVA: state the hypotheses, verify that the conditions for ANOVA are met, look at the F statistic and its P-value, and state a conclusion.

One-Way Analysis of Variance

Analysis of Variance for Score

Source	DF	SS	MS	F	P
Group	2	20.58	10.29	1.13	0.329
Error	63	572.45	9.09		
Total	65	593.03			

Pooled StDev = 3.014

FIGURE 14.8 Analysis of variance output from Minitab for the pretest scores in the new-product evaluation study.

14.6 **An alternative Normal quantile plot.** Figure 14.6 displays separate Normal quantile plots for the three groups. An alternative procedure is to make one Normal quantile plot using the *residuals* $e_{ij} = \overline{x}_{ij} - \overline{x}_i$ for all three groups together. Make this plot and summarize what it shows.

The ANOVA table

Software ANOVA output contains more than simply the test statistic. The additional information shows, among other things, where the test statistic comes from.

ANOVA table

The information in an analysis of variance is organized in an **ANOVA table.** In the software output in Figure 14.8, the columns of this table are labeled Source, DF, SS, MS, F, and P. The rows are labeled Group, Error, and Total. These are the three sources of variation in the one-way ANOVA. ("Group" was the name used in entering the data to distinguish the three treatment groups.)

variation among groups

variation within groups

The Group row in the table corresponds to the FIT term in our DATA = FIT + RESIDUAL way of thinking. It gives information related to the variation **among** group means. The ANOVA model allows the groups to have different means. The Error row in the table corresponds to the RESIDUAL term in DATA = FIT + RESIDUAL. It gives information related to the variation **within** groups. The term "error" is most appropriate for experiments in the physical sciences, where the observations within a group differ because of measurement error. In business and the biological and social sciences, on the other hand, the within-group variation is often due to the fact that not all firms or plants or people are the same. This sort of variation is not due to errors and is better described as "residual." Finally, the Total row in the table corresponds to the DATA term in our DATA = FIT + RESIDUAL framework.

For analysis of variance, the idea that

$$\text{DATA} = \text{FIT} + \text{RESIDUAL}$$

translates into an actual equation:

$$\text{total variation} = \text{variation among groups} + \text{variation within groups}$$

sums of squares

The ANOVA idea is to break the total variation in the responses into two parts: the variation due to differences among the group means and that due to differences within groups. Variation is expressed by **sums of squares.** We use SSG, SSE, and SST for the sums of squares for groups, error, and total. Each sum of squares is the sum of the squares of a set of deviations that expresses a source of variation. SST is the sum of squares of $x_{ij} - \overline{x}$, which measure variation of the responses around their overall mean. Variation of the group means around the overall mean $\overline{x}_i - \overline{x}$ is measured by SSG. Finally,

SSE is the sum of squares of the deviations $x_{ij} - \overline{x}_i$ of each observation from its group mean. It is always true that SST = SSG + SSE. This is the algebraic version of the ANOVA idea: total variation is the sum of among-group variation and within-group variation.

EXAMPLE 14.8 **Sums of squares for the three sources of variation**

CASE 14.1 

The SS column in Figure 14.8 gives the values for the three sums of squares. The output gives many more digits than we need. Rounded-off values are SSG = 21, SSE = 572, and SST = 593. In this example it appears that most of the variation is coming from Error, that is, from within groups.

degrees of freedom

Associated with each sum of squares is a quantity called the **degrees of freedom.** Because SST measures the variation of all N observations around the overall mean, its degrees of freedom are DFT $= N - 1$, the degrees of freedom for the sample variance of the N responses. Similarly, because SSG measures the variation of the I sample means around the overall mean, its degrees of freedom are DFG $= I - 1$. Finally, SSE is the sum of squares of the deviations $x_{ij} - \overline{x}_i$. Here we have N observations being compared with I sample means and DFE $= N - I$.

EXAMPLE 14.9 **Degrees of freedom for the three sources**

CASE 14.1 

The DF column in Figure 14.8 gives the values for the three degrees of freedom. These values are DFT = 65, DFG = 2, and DFE = 63.

mean square

For each source of variation, the **mean square** is the sum of squares divided by the degrees of freedom. Generally, the ANOVA table includes mean squares only for the first two sources of variation.

EXAMPLE 14.10 **Mean squares for the three sources**

CASE 14.1 

The MS column in Figure 14.8 gives values for two of the mean squares. These values are MSG = 10.29 and MSE = 9.09.

The mean square corresponding to the total source is the sample variance that we would calculate assuming that we have one sample from a single population—that is, assuming that the means of the three groups are the same.

SUMS OF SQUARES, DEGREES OF FREEDOM, AND MEAN SQUARES

Sums of squares represent variation present in the data. They are calculated by summing squared deviations. In one-way ANOVA there are three **sources of variation:** groups, error, and total. The sums of squares are related by the formula

$$\text{SST} = \text{SSG} + \text{SSE}$$

Thus, the total variation is composed of two parts, one due to groups and one due to "error" (variation within groups).

Degrees of freedom are related to the deviations that are used in the sums of squares. The degrees of freedom are related in the same way as the sums of squares:

$$\text{DFT} = \text{DFG} + \text{DFE}$$

To calculate each **mean square,** divide the corresponding sum of squares by its degrees of freedom.

All of these exercises use the output in Figure 14.8 or the data in Table 14.1.

14.7 **Verify that the SS add.** Verify that the sums of squares add, SST = SSG + SSE.

14.8 **Check the DF.** What are I and N for these data? Verify that the DF entries in the output are DFG = $I - 1$, DFE = $N - I$, and DFT = $N - 1$.

14.9 **Do the DF add?** Verify that degrees of freedom add in the same way that the sums of squares add. That is, DFT = DFG + DFE.

14.10 **Check the MS.** Verify that each mean square in the output is the corresponding sum of squares divided by its degrees of freedom.

14.11 **Total mean square.** The output does not give the total mean square MST = SST/DFT. Calculate this quantity. Then find the mean and variance of all 66 observations in Table 14.1 and verify that MST is the variance of all the responses.

The *F* test

The ANOVA table also reports the pooled standard error s_p. It is always true that

$$s_p^2 = \text{MSE} = \frac{\text{SSE}}{\text{DFE}}$$

This fact reinforces the idea that MSE measures variation within groups. From the output in Figure 14.8, we can calculate the variance $s_p^2 = 3.014^2 = 9.08$. This is the same as MSE up to roundoff error.

The F statistic is the final entry in the ANOVA table. If H_0 is true, there are no differences among the group means. Then MSG will reflect only chance variation and we expect MSG to be about the same as MSE. The

F statistic simply compares these two mean squares, F = MSG/MSE. This statistic is near 1 if H_0 is true and tends to be larger if H_a is true. In our example, MSG = 10.29 and MSE = 9.09, so the ANOVA F statistic is

$$F = \frac{\text{MSG}}{\text{MSE}} = \frac{10.29}{9.09} = 1.13$$

When H_0 is true, the F statistic has an F distribution that depends upon two numbers: the *degrees of freedom for the numerator* and the *degrees of freedom for the denominator*. These degrees of freedom are those associated with the mean squares in the numerator and denominator of the F statistic. For one-way ANOVA, the degrees of freedom for the numerator are DFG = $I - 1$ and the degrees of freedom for the denominator are DFE = $N - I$. We use the notation $F(I - 1, N - I)$ for this distribution.

THE ANOVA *F* TEST

To test the null hypothesis in a one-way ANOVA, calculate the **F statistic**

$$F = \frac{\text{MSG}}{\text{MSE}}$$

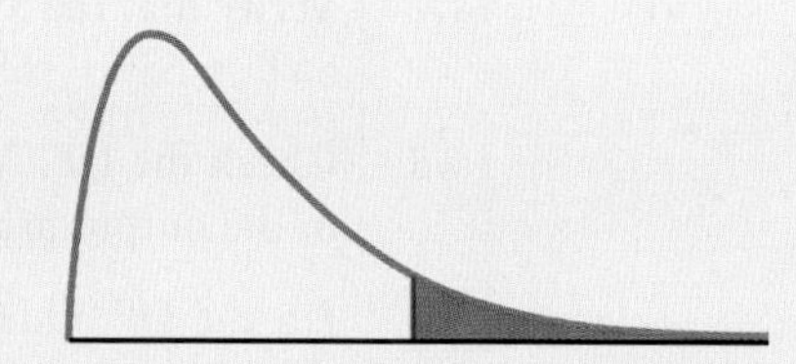

When H_0 is true, the F statistic has the $F(I - 1, N - I)$ distribution. When H_a is true, the F statistic tends to be large. We reject H_0 in favor of H_a if the F statistic is sufficiently large.

The ***P*-value** of the F test is the probability that a random variable having the $F(I - 1, N - I)$ distribution is greater than or equal to the calculated value of the F statistic.

Tables of F critical values are available for use when software does not give the P-value. Table E in the back of the book contains the F critical values for probabilities p = 0.100, 0.050, 0.025, 0.010, and 0.001. For one-way ANOVA we use critical values from the table corresponding to $I - 1$ degrees of freedom in the numerator and $N - I$ degrees of freedom in the denominator. We have already seen several examples where the F statistic and its P-value were used to choose between H_0 and H_a.

EXAMPLE 14.11 **The mean GPAs are not identical**

In the study of computer science students in Example 14.3, F = 10.35. There are three populations, so the degrees of freedom in the numerator are

p	Critical value
0.100	2.33
0.050	3.04
0.025	3.76
0.010	4.71
0.001	7.15

DFG $= I - 1 = 2$. The degrees of freedom in the denominator are DFE $= N - I = 256 - 3 = 253$. In Table E first find the column corresponding to 2 degrees of freedom in the numerator. For the degrees of freedom in the denominator, there are entries for 200 and 1000. These entries are very close. To be conservative we use critical values corresponding to 200 degrees of freedom in the denominator since these are slightly larger. Because 10.35 is larger than all of the tabulated values, we reject H_0 and conclude that the differences in means are statistically significant, with $P < 0.001$.

EXAMPLE 14.12

Mean pretest scores do not differ significantly

p	Critical value
0.100	2.39
0.050	3.15
0.025	3.93
0.010	4.98
0.001	7.77

For comparing the three teaching methods in Case 14.1, $F = 1.13$. Because there are three populations being compared, the degrees of freedom in the numerator are DFG $= I - 1 = 2$, and the degrees of freedom in the denominator are DFE $= N - I = 66 - 3 = 63$. Table E has entries for 2 degrees of freedom in the numerator. There is no entry for 63 degrees of freedom in the denominator, so we use the slightly larger values corresponding to 60. Because 1.13 is less than any of the tabulated values, we do not have strong evidence in the data against the hypothesis of equal means. We conclude that the differences in sample means are not significant at the 10% level, $P > 0.1$. Statistical software (Figure 14.8) gives the exact value $P = 0.329$.

Remember that the F test is always one-sided because any differences among the group means tend to make F large. The ANOVA F test shares the robustness of the two-sample t test. It is relatively insensitive to moderate non-Normality and unequal variances, especially when the sample sizes are similar.

APPLY YOUR KNOWLEDGE

14.12 Pool the variances. Find the variances for the three samples by squaring the standard deviations given in the output in Figure 14.7. Then pool the variances using the formula given on page 14-13. Verify that this calculation gives the MSE entry in Figure 14.8.

14.13 Use Table E. An ANOVA is run to compare 4 groups. There are 6 observations in each group.

(a) Give the degrees of freedom for the ANOVA F statistic.

(b) How large would this statistic need to be to have a P-value less than 0.05?

14.14 Suppose we had more data. Refer to the previous exercise. Suppose that we are still interested in comparing the 4 groups, but we obtain data on 26 subjects per group. Answer parts (a) and (b) of the previous exercise. Can you explain why the answer to part (b) is smaller than what you found for part (b) of the previous exercise?

Using software

The following display shows the general form of a one-way ANOVA table with the F statistic. The formulas in the sum of squares column can be used for calculations in small problems. There are other formulas that are more efficient for hand or calculator use, but ANOVA calculations are usually done by computer software.

Source	Degrees of freedom	Sum of squares	Mean square	F
Groups	$I - 1$	$\sum_{\text{groups}} n_i(\bar{x}_i - \bar{x})^2$	SSG/DFG	MSG/MSE
Error	$N - I$	$\sum_{\text{groups}}(n_i - 1)s_i^2$	SSE/DFE	
Total	$N - 1$	$\sum_{\text{obs}}(x_{ij} - \bar{x})^2$	SST/DFT	

You should now be able to extract all the ANOVA information we have discussed from output from almost any statistical software. Here is a final example on which to practice this skill.

EXAMPLE 14.13 **Advertising and the quality of a product**

Research suggests that customers think that a product is of high quality if it is heavily advertised. An experiment designed to explore this idea collected quality ratings (on a 1 to 7 scale) of a new line of take-home refrigerated entrées based on reading a magazine ad. Three groups were compared. The first group's ad included information that would undermine (U) the expected positive association between quality and advertising; the second group's ad contained information that would affirm (A) the association; and the third group was a control (C).[4] The data are given in Table 14.2. Outputs from SAS, Excel, and Minitab appear in Figure 14.9.

We recommend the *One-Way ANOVA* applet available on the Web site `www.whfreeman.com/pbs` as an excellent way to see how the value of the F statistic and its P-value depend on both the variability of the data within the groups and the differences among the group means. Exercises 14.16 and 14.17 make use of this applet.

TABLE 14.2 Quality ratings in three groups

Group	Quality ratings
Undermine	6 5 5 5 4 5 4 6 5 5 5 5 3 3 5 4 5 5 5 4
(n = 55)	5 4 4 5 4 4 4 5 5 5 4 5 5 5 4 4 5 5 4 5
	5 4 4 5 4 3 4 5 5 5 3 4 4 4 4
Affirm	4 6 4 6 5 5 5 6 4 5 5 5 4 6 6 5 5 7 4 6
(n = 36)	6 4 5 4 5 5 6 4 5 5 4 6 4 6 5 5
Control	5 4 5 6 5 7 5 6 7 5 7 5 4 5 4 4 6 6 5 6
(n = 36)	5 5 4 5 5 6 6 6 5 6 6 7 6 6 5 5

SAS

```
General Linear Models Procedure

Dependent Variable: QUALITY
                                Sum of              Mean
Source                  DF     Squares            Square   F Value      Pr > F
Model                    2   18.828255          9.414127     15,28      0.0001
Error                  124   76.384343          0.616003
Corrected Total        126   95.212598

                  R-Square        C.V.          Root MSE          QUALITY Mean
                  0.197750    15.94832            0.7849                4.9213

Level of          -----------QUALITY-----------
             GROUP       N          Mean                  SD

             affirm     36    5.05555556          0.82615960
             control    36    5.41666667          0.87423436
             under      55    4.50909091          0.69048365
```

Excel

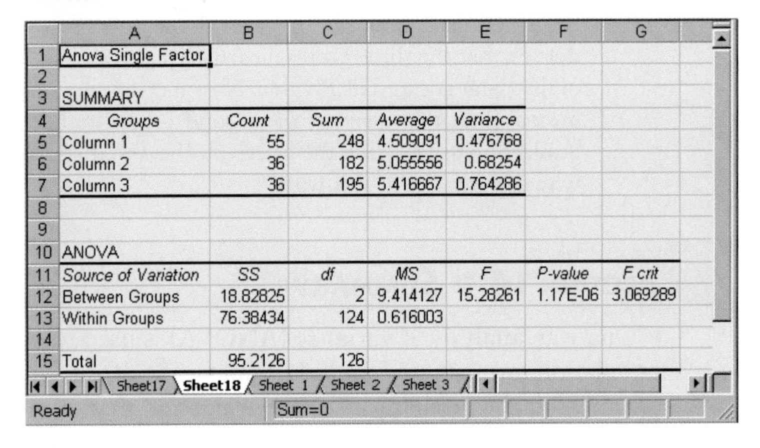

	A	B	C	D	E	F	G
1	Anova Single Factor						
2							
3	SUMMARY						
4	*Groups*	*Count*	*Sum*	*Average*	*Variance*		
5	Column 1	55	248	4.509091	0.476768		
6	Column 2	36	182	5.055556	0.68254		
7	Column 3	36	195	5.416667	0.764286		
8							
9							
10	ANOVA						
11	*Source of Variation*	*SS*	*df*	*MS*	*F*	*P-value*	*F crit*
12	Between Groups	18.82825	2	9.414127	15.28261	1.17E-06	3.069289
13	Within Groups	76.38434	124	0.616003			
14							
15	Total	95.2126	126				

Sheet17 Sheet18 Sheet 1 Sheet 2 Sheet 3

Ready Sum=0

FIGURE 14.9(a) SAS and Excel output for the advertising study.

14.15 Compare software. The pooled standard error for the data in Table 14.2 is $s_p = 0.7849$. Look at the software output in Figure 14.9.

(a) Explain to someone new to ANOVA why SAS labels this quantity as "Root MSE."

(b) Excel does not report s_p. How can you find its value from Excel output?

14.16 The effect of within-group variation. Go to the *One-Way ANOVA* applet. In the applet display, the black dots are the mean responses in three treatment groups. Move these up and down until you get a configuration with *P*-value about 0.01. Note the value of the *F* statistic. Now increase the variation within the groups without changing their means by dragging the mark on the pooled standard error scale to the right. Describe what happens to the *F* statistic and the *P*-value. Explain why this happens.

Minitab

```
One-Way Analysis of Variance

Analysis of Variance
Source     DF        SS        MS        F        P
Factor      2    18.828     9.414    15.28   0.0000
Error     124    76.384     0.616
Total     126    95.123
                                  Individual 95% CIs For Mean
                                  Based on Pooled StDev
Level       N      Mean     StDev  ---+---------+---------+---------+---
Undermin   55    4.5091    0.6905  (-----*----)
Affirm     36    5.0556    0.8262             (-----*------)
Control    36    5.4167    0.8742                      (-----*------)
                                   ---+---------+---------+---------+---
Pooled StDev =   0.7849              4.40      4.80      5.20      5.60
```

FIGURE 14.9(b) Minitab output for the advertising study.

14.17 **The effect of among-group variation.** Go to the *One-Way ANOVA* applet. Set the pooled standard error near the middle of its scale and drag the black dots so that the three group means are approximately equal. Note the value of the F statistic and its P-value. Now increase the variation among the group means: drag the mean of the second group up and the mean of the third group down. Describe the effect on the F statistic and its P-value. Explain why they change in this way.

SECTION 14.1 SUMMARY

■ **One-way analysis of variance (ANOVA)** is used to compare several population means based on independent SRSs from each population. We assume that the populations are Normal and that, although they may have different means, they have the same standard deviation.

■ To do an analysis of variance, first examine the data. Side-by-side boxplots give an overview. Examine Normal quantile plots (either for each group separately or for the residuals) to detect outliers or extreme deviations from Normality. Compute the ratio of the largest to the smallest sample standard deviation. If this ratio is less than 2 and the Normal quantile plots are satisfactory, ANOVA can be performed.

■ The **null hypothesis** is that the population means are *all equal*. The **alternative hypothesis** is true if there are *any* differences among the population means.

■ ANOVA is based on separating the total variation observed in the data into two parts: variation **among group means** and variation **within groups.** If the variation among groups is large relative to the variation within groups, we have evidence against the null hypothesis.

- An **analysis of variance table** organizes the ANOVA calculations. **Degrees of freedom, sums of squares, and mean squares** appear in the table. The ***F* statistic** and its ***P*-value** are used to test the null hypothesis.

14.2 Comparing Group Means

The ANOVA F test gives a general answer to a general question: Are the differences among observed group means significant? Unfortunately, a small P-value simply tells us that the group means are not all the same. It does not tell us specifically which means differ from each other. Plotting and inspecting the means give us some indication of where the differences lie, but we would like to supplement inspection with formal inference. This section presents two approaches to the task of comparing group means.

Contrasts

The preferred approach is to pose specific questions regarding comparisons among the means before the data are collected. We can answer specific questions of this kind and attach a level of confidence to the answers we give. We now explore these ideas through an example.

CASE 14.2

EVALUATION OF THE NEW PRODUCT

Case 14.1 introduces a randomized comparative experiment to compare three methods for teaching reading. We analyzed the pretest scores and found no reason to reject H_0 that the three groups had similar population means on this measure. This was the desired outcome. We now turn to the response variable, a measure of reading comprehension called COMP that was measured by a test taken after the instruction was completed.

The Basal method is the standard method commonly used in schools. Your company will market materials based on two innovative methods, called DRTA and Strat. The DRTA and Strat methods are not identical, but both involve teaching the students to use similar comprehension strategies in their reading. Can you claim that the new methods are superior to Basal?

We can compare new with standard by posing and answering specific questions about the mean responses. First, here is the basic ANOVA.

EXAMPLE 14.14 **Are the comprehension scores different?**

CASE 14.2

Figure 14.10 gives the summary statistics for COMP computed by SPSS. This software uses only numeric values for the factor, so we coded the groups as 1 for Basal, 2 for DRTA, and 3 for Strat. Side-by-side boxplots appear in Figure 14.11,

COMP			Descriptives
	N	Mean	Std. Deviation
1.00	22	41.0455	5.63558
2.00	22	46.7273	7.38842
3.00	22	44.2727	5.76675
Total	66	44.0152	6.64366

FIGURE 14.10 Summary statistics from SPSS for the comprehension scores in the three groups in the new-product evaluation study.

and Figure 14.12 plots the group means. The ANOVA results generated by SPSS are given in Figure 14.13, and a Normal quantile plot of the residuals appears in Figure 14.14.

The ANOVA null hypothesis is

$$H_0\colon \mu_B = \mu_D = \mu_S$$

where the subscripts correspond to the group labels Basal, DRTA, and Strat. Figure 14.13 shows that $F = 4.48$ with degrees of freedom 2 and 63. The P-value is 0.015. We have good evidence against H_0.

What can the researchers conclude from this analysis? The alternative hypothesis is true if $\mu_B \neq \mu_D$ or if $\mu_B \neq \mu_S$ or if $\mu_D \neq \mu_S$ or if any combination of these statements is true. We would like to be more specific.

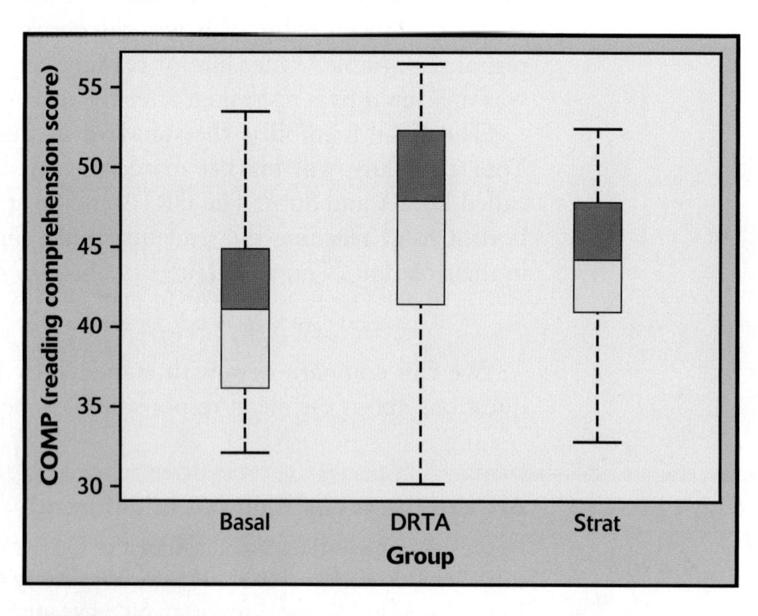

FIGURE 14.11 Side-by-side boxplots of the comprehension scores in the new-product evaluation study.

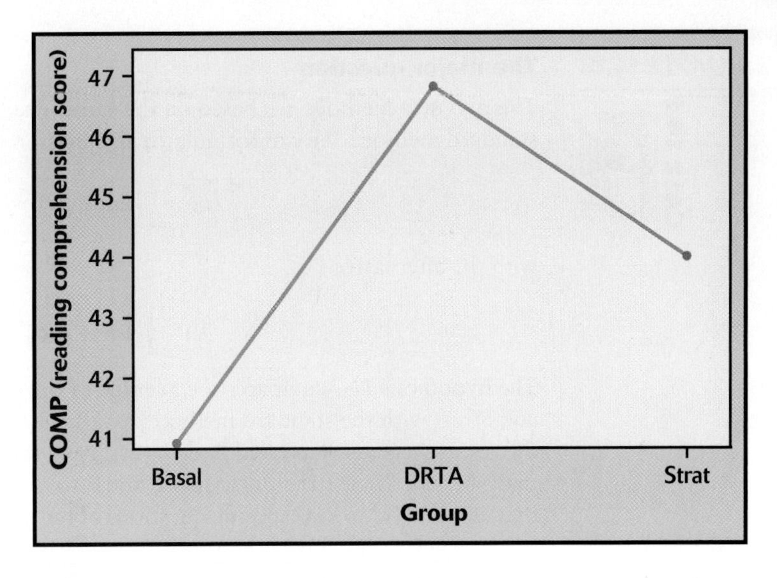

FIGURE 14.12 Comprehension score group means in the new-product evaluation study.

ANOVA
COMP

	Sum of Squares	df	Mean Square	F	Sig.
Between Groups	357.303	2	178.652	4.481	0.015
Within Groups	2511.682	63	39.868		
Total	2868.985	65			

FIGURE 14.13 Analysis of variance output from SPSS for the comprehension scores in the new-product evaluation study.

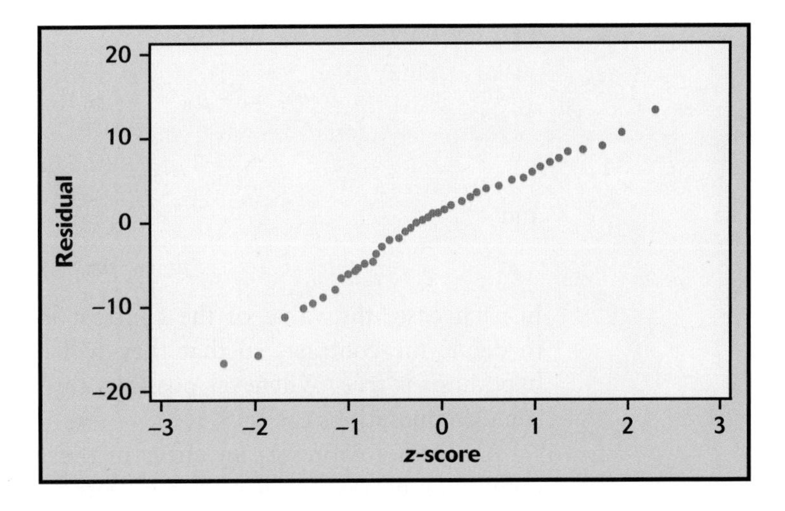

FIGURE 14.14 Normal quantile plot of the residuals for the comprehension scores in the new-product evaluation study.

EXAMPLE 14.15 CASE 14.2

The major question

The two new methods are based on the same idea. Are they superior to the standard method? We can formulate this question as the null hypothesis

$$H_{01}: \frac{1}{2}(\mu_D + \mu_S) = \mu_B$$

with the alternative

$$H_{a1}: \frac{1}{2}(\mu_D + \mu_S) > \mu_B$$

The hypothesis H_{01} compares the average of the two innovative methods (DRTA and Strat) with the standard method (Basal). The alternative is one-sided because the researchers are interested in demonstrating that the new methods are better than the old. We use the subscripts 1 and 2 to distinguish two sets of hypotheses that correspond to two specific questions about the means.

EXAMPLE 14.16 CASE 14.2

A secondary question

A secondary question involves a comparison of the two new methods. We formulate this as the hypothesis that the methods DRTA and Strat are equally effective,

$$H_{02}: \mu_D = \mu_S$$

versus the alternative

$$H_{a2}: \mu_D \neq \mu_S$$

contrast

ψ

Each of H_{01} and H_{02} says that a combination of population means is 0. These combinations of means are called **contrasts.** We use ψ, the Greek letter psi, for contrasts among population means. The two contrasts that arise from our two null hypotheses are

$$\psi_1 = -\mu_B + \frac{1}{2}(\mu_D + \mu_S)$$
$$= (-1)\mu_B + (0.5)\mu_D + (0.5)\mu_S$$

and

$$\psi_2 = \mu_D - \mu_S$$

In each case, the value of the contrast is 0 when H_0 is true. We chose to define the contrasts so that they will be positive when the alternative hypothesis is true. Whenever possible, this is a good idea because it makes some computations easier.

sample contrast

A contrast expresses an effect in the population as a combination of population means. To estimate the contrast, form the corresponding **sample contrast** by using sample means in place of population means. Under the ANOVA assumptions, a sample contrast is a linear combination of independent Normal variables and therefore has a Normal distribution. We can obtain the standard error of a contrast by using the rules for variances given in Section 4.3. Inference is based on t statistics. Here are the details.

CONTRASTS

A **contrast** is a combination of population means of the form

$$\psi = \sum a_i \mu_i$$

where the coefficients a_i have sum 0. The corresponding **sample contrast** is the same combination of sample means,

$$c = \sum a_i \bar{x}_i$$

The **standard error of c** is

$$\text{SE}_c = s_p \sqrt{\sum \frac{a_i^2}{n_i}}$$

To test the null hypothesis H_0: $\psi = 0$, use the ***t* statistic**

$$t = \frac{c}{\text{SE}_c}$$

with degrees of freedom DFE that are associated with s_p. The alternative hypothesis can be one-sided or two-sided.

A **level C confidence interval for ψ** is

$$c \pm t^* \text{SE}_c$$

where t^* is the value for the t(DFE) density curve with area C between $-t^*$ and t^*.

Because each $\bar{x}_i$ estimates the corresponding μ_i, the addition rule for means tells us that the mean μ_c of the sample contrast c is ψ. In other words, c is an unbiased estimator of ψ. Testing the hypothesis that a contrast is 0 assesses the significance of the effect measured by the contrast. It is often more informative to estimate the size of the effect using a confidence interval for the population contrast.

EXAMPLE 14.17 **The coefficients for the contrasts**

CASE 14.2

In our example the coefficients in the contrasts are $a_1 = -1, a_2 = 0.5, a_3 = 0.5$ for ψ_1 and $a_1 = 0, a_2 = 1, a_3 = -1$ for ψ_2, where the subscripts 1, 2, and 3 correspond to B, D, and S. In each case the sum of the a_i is 0.

We look at inference for each of these contrasts in turn.

EXAMPLE 14.18 **Are the new methods better?**

CASE 14.2

The sample contrast that estimates ψ_1 is

$$c_1 = \bar{x}_B + \frac{1}{2}(\bar{x}_D + \bar{x}_S)$$

$$= -41.05 + \frac{1}{2}(46.73 + 44.27) = 4.45$$

with standard error

$$SE_{c_1} = 6.314\sqrt{\frac{(-1)^2}{22} + \frac{(0.5)^2}{22} + \frac{(0.5)^2}{22}}$$

$$= 1.65$$

The t statistic for testing H_{01}: $\psi_1 = 0$ versus H_{a1}: $\psi_1 > 0$ is

$$t = \frac{c_1}{SE_{c_1}}$$

$$= \frac{4.45}{1.65} = 2.70$$

Because s_p has 63 degrees of freedom, software using the $t(63)$ distribution gives the one-sided P-value as 0.0044. If we used Table D, we would conclude that $P < 0.005$. The P-value is small, so there is strong evidence against H_{01}. The researchers have shown that the new methods produce higher mean scores than the old. The size of the improvement can be described by a confidence interval. To find the 95% confidence interval for ψ_1, we combine the estimate with its margin of error:

$$c_1 \pm t^* SE_{c_1} = 4.45 \pm (2.00)(1.65)$$

$$= 4.45 \pm 3.30$$

The interval is (1.15, 7.75). We are 95% confident that the mean improvement obtained by using one of the innovative methods rather than the old method is between 1.15 and 7.75 points.

EXAMPLE 14.19 **Comparing the two new methods**

The second sample contrast, which compares the two new methods, is

$$c_2 = 46.73 - 44.27$$

$$= 2.46$$

with standard error

$$SE_{c_2} = 6.314\sqrt{\frac{(1)^2}{22} + \frac{(-1)^2}{22}}$$

$$= 1.90$$

The t statistic for assessing the significance of this contrast is

$$t = \frac{2.46}{1.90} = 1.29$$

The P-value for the two-sided alternative is 0.2020. We conclude that either the two new methods have the same population means or the sample sizes are not sufficiently large to distinguish them. A confidence interval helps clarify this statement. To find the 95% confidence interval for ψ_2, we combine the estimate with its margin of error:

$$\begin{aligned} c_2 \pm t^* \text{SE}_{c_2} &= 2.46 \pm (2.00)(1.90) \\ &= 2.46 \pm 3.80 \end{aligned}$$

The interval is $(-1.34, 6.26)$. With 95% confidence we state that the difference between the population means for the two new methods is between -1.34 and 6.26.

EXAMPLE 14.20 **Using software**

CASE 14.2

Figure 14.15 displays the SPSS output for the analysis of these contrasts. The column labeled "t" gives the t statistics 2.702 and 1.289 for our two contrasts. The degrees of freedom appear in the column labeled "df" and are 63 for each t. The P-values are given in the column with the label "Sig. (2-tailed)." These are correct for two-sided alternative hypotheses. The values are 0.009 and 0.202. To convert the computer-generated results to apply to our one-sided alternative concerning ψ_1, simply divide the reported P-value by 2 after checking that the value of c is in the direction of H_a (that is, that c is positive).

Some statistical software packages report the test statistics associated with contrasts as F statistics rather than t statistics. These F statistics are the squares of the t statistics described above. The associated P-values are for the two-sided alternatives.

Questions about population means are expressed as hypotheses about contrasts. A contrast should express a specific question that we have in mind when designing the study. When contrasts are formulated before seeing the data, *inference about contrasts is valid whether or not the ANOVA H_0 of*

Contrast Coefficients

	GROUPN		
Contrast	1.00	2.00	3.00
1	−1	0.5	0.5
2	0	1	−1

Contrast Tests

	Contrast	Value of Contrast	Std. Error	t	df	Sig. (2-tailed)
COMP	1	4.4545	1.64872	2.702	63	0.009
	2	2.4545	1.90378	1.289	63	0.202

FIGURE 14.15 SPSS output for contrasts for the comprehension scores in the new-product evaluation study.

equality of means is rejected. Because the F test answers a very general question, it is less powerful than tests for contrasts designed to answer specific questions. Specifying the important questions before the analysis is undertaken enables us to use this powerful statistical technique.

APPLY YOUR KNOWLEDGE

14.18 **Define a contrast.** An ANOVA was run with five groups. Give the coefficients for the contrast that compares the average of the means of the first two groups with the average of the means of the last two groups.

14.19 **Find the standard error.** Refer to the previous exercise. Suppose that there are 25 observations in each group and that $s_p = 10$. Find the standard error for the contrast.

14.20 **Is the contrast significant?** Suppose that the average of the first two groups minus the average of the last two groups is 15. State an appropriate null hypothesis for this comparison, find the test statistic with its degrees of freedom, and report the result.

14.21 **Give the confidence interval.** Give a 95% confidence interval for the difference between the average of the means of the first two groups and the average of the means of the last two groups.

Multiple comparisons

In many studies, specific questions cannot be formulated in advance of the analysis. If H_0 is not rejected, we conclude that the population means are indistinguishable on the basis of the data given. On the other hand, if H_0 is rejected, we would like to know which pairs of means differ. **Multiple comparisons** methods address this issue. It is important to keep in mind that multiple comparisons methods are used only *after rejecting* the ANOVA H_0.

multiple comparisons

Return once more to the reading comprehension study described in Case 14.2. We found in Example 14.14 that the means were not all the same ($F = 4.48$, df = 2 and 63, $P = 0.015$).

EXAMPLE 14.21 **A t statistic to compare two means**

CASE 14.2

There are three pairs of population means. We can compare groups 1 and 2, groups 1 and 3, and groups 2 and 3. For each of these pairs, we can write a t statistic for the difference in means. To compare Basal with DRTA (1 with 2), we compute

$$t_{12} = \frac{\bar{x}_1 - \bar{x}_2}{s_p\sqrt{\frac{1}{n_1} + \frac{1}{n_2}}}$$

$$= \frac{41.05 - 46.73}{6.31\sqrt{\frac{1}{22} + \frac{1}{22}}}$$

$$= -2.99$$

The subscripts on t specify which groups are compared.

14.22 Compare Basal with Strat. Verify that $t_{13} = -1.69$.

14.23 Compare DRTA with Strat. Verify that $t_{23} = 1.29$. (This is the same t that we used for the contrast $\psi_2 = \mu_2 - \mu_3$ in Example 14.19.)

These t statistics are very similar to the pooled two-sample t statistic for comparing two population means, described in Chapter 7 (page 475). The difference is that we now have more than two populations, so each statistic uses the pooled estimator s_p from all groups rather than the pooled estimator from just the two groups being compared. This additional information about the common σ increases the power of the tests. The degrees of freedom for all of these statistics are DFE = 63, those associated with s_p.

Because we do not have any specific ordering of the means in mind as an alternative to equality, we must use a two-sided approach to the problem of deciding which pairs of means are significantly different.

MULTIPLE COMPARISONS

To perform a **multiple comparisons procedure**, compute ***t* statistics** for all pairs of means using the formula

$$t_{ij} = \frac{\overline{x}_i - \overline{x}_j}{s_p\sqrt{\frac{1}{n_i} + \frac{1}{n_j}}}$$

If

$$|t_{ij}| \geq t^{**}$$

we declare that the population means μ_i and μ_j are different. Otherwise, we conclude that the data do not distinguish between them. The value of t^{**} depends upon which multiple comparisons procedure we choose.

One obvious choice for t^{**} is the upper $\alpha/2$ critical value for the t(DFE) distribution. This choice simply carries out as many separate significance tests of fixed level α as there are pairs of means to be compared. The procedure based on this choice is called the **least-significant differences** method, or simply LSD. LSD has some undesirable properties, particularly if the number of means being compared is large. Suppose, for example, that there are $I = 20$ groups and we use LSD with $\alpha = 0.05$. There are 190 different pairs of means. If we perform 190 t tests, each with an error rate of 5%, our overall error rate will be unacceptably large. We expect about 5% of the 190 to be significant even if the corresponding population means are the same. Since 5% of 190 is 9.5, we expect 9 or 10 false rejections.

LSD method

The LSD procedure fixes the probability of a false rejection for each single pair of means being compared. It does not control the overall probability of *some* false rejection among all pairs. Other choices of t^{**} control possible errors in other ways. The choice of t^{**} is therefore a complex problem and a detailed discussion of it is beyond the scope of this text. Many choices for t^{**} are used in practice. One major statistical package allows selection from a list of over a dozen choices.

Bonferroni method

We will discuss only one of these, called the **Bonferroni method.** Use of this procedure with $\alpha = 0.05$, for example, guarantees that the probability of *any* false rejection among all comparisons made is no greater than 0.05. This is much stronger protection than controlling the probability of a false rejection at 0.05 for *each separate* comparison.

EXAMPLE 14.22 **Which means differ?**

We apply the Bonferroni multiple comparisons procedure with $\alpha = 0.05$ to the data from the new-product evaluation study in Example 14.14. The value of t^{**} for this procedure (from software or special tables) is 2.46. The t statistic for comparing Basal with DRTA is $t_{12} = -2.99$. Because $|-2.99|$ is greater than 2.46, the value of t^{**}, we conclude that the DRTA method produces higher reading comprehension scores than Basal.

APPLY YOUR KNOWLEDGE

14.24 **Compare Basal with Strat.** The test statistic for comparing Basal with Strat is $t_{13} = -1.69$. For the Bonferroni multiple comparisons procedure with $\alpha = 0.05$, do you reject the null hypothesis that the population means for these two groups are different?

14.25 **Compare DRTA with Strat.** Answer the same question for the comparison of DRTA with Strat using the calculated value $t_{23} = 1.29$.

Usually we use software to perform the multiple comparisons procedure. The formats differ from package to package but they all give the same basic information.

EXAMPLE 14.23 **Computer output for multiple comparisons**

The output from SPSS for Bonferroni comparisons appears in Figure 14.16. The first line of numbers gives the results for comparing Basal with DRTA, groups 1 and 2. The difference between the means is given as -5.6818 with a standard error of 1.90378. The P-value for the comparison is given under the heading "Sig." The value is 0.012. Therefore, we could declare the means for Basal and DRTA to be different according to the Bonferroni procedure as long as we are using a value of α that is less than or equal to 0.012. In particular, these groups are significantly different at the *overall* $\alpha = 0.05$ level. The last two entries in the row give the Bonferroni 95% confidence interval. We will discuss this later.

SPSS does not give the values of the t statistics for multiple comparisons. To compute them, simply divide the difference in the means by the standard error. For comparing Basal with DRTA, we have as before

$$t_{12} = \frac{-5.681}{1.90} = -2.99$$

Post Hoc Tests

Multiple Comparisons
Dependent Variable: COMP
Bonferroni

(I) GROUP	(J) GROUP	Mean Difference (I-J)	Std. Error	Sig.	95% Confidence Interval Lower Bound	Upper Bound
1.00	2.00	−5.6818	1.90378	0.012	−10.3643	−0.9993
	3.00	−3.2273	1.90378	0.285	−7.9098	1.4552
2.00	1.00	5.6818	1.90378	0.012	0.9993	10.3643
	3.00	2.4545	1.90378	0.606	−2.2280	7.1370
3.00	1.00	3.2273	1.90378	0.285	−1.4552	7.9098
	2.00	−2.4545	1.90378	0.606	−7.1370	2.2280

* The mean difference is significant at the 0.05 level.

FIGURE 14.16 SPSS output for Bonferroni multiple comparisons and simultaneous confidence intervals for the comprehension scores in the new-product evaluation study.

14.26 **Compare Basal with Strat.** Use the difference in means and the standard error reported in the output of Figure 14.16 to verify that $t_{13} = -1.69$.

14.27 **Compare DRTA with Strat.** Use the difference in means and the standard error reported in the output of Figure 14.16 to verify that $t_{23} = 1.29$.

When there are many groups, the many results of multiple comparisons are difficult to describe. Here is one common format.

EXAMPLE 14.24 **Displaying multiple comparisons results**

Here is a table of the means and standard deviations for the three treatment groups. To report the results of multiple comparisons, use letters to label the means of pairs of groups that do *not* differ at the overall 0.05 significance level.

Group	Mean	SD	n
Basal	41.05^{A}	2.97	22
DRTA	46.73^{B}	2.65	22
Strat	$44.27^{A,B}$	3.34	22

Label "A" shows that Basal and Strat do not differ. Label "B" shows that DRTA and Strat do not differ. Because Basal and DRTA do not have a common label, they do differ.

The display in Example 14.24 shows that at the overall 0.05 significance level Basal does not differ from Strat and Strat does not differ from DRTA, yet Basal does differ from DRTA. These conclusions appear to be illogical. If μ_1 is the same as μ_3, and μ_3 is the same as μ_2, doesn't it follow that μ_1 is the same as μ_2? Logically, the answer must be yes.

This apparent contradiction points out dramatically the nature of the conclusions of tests of significance. A careful statement would say that we found significant evidence that Basal differs from DRTA and failed to find evidence that Basal differs from Strat or that Strat differs from DRTA. Failing to find strong enough evidence that two means differ doesn't say that they are equal. It is very unlikely that any two methods of teaching reading comprehension would give *exactly* the same population means, but the data can fail to provide good evidence of a difference. This is particularly true in multiple comparisons methods such as Bonferroni that use a single α for an entire set of comparisons.

APPLY YOUR KNOWLEDGE

14.28 **Which means differ significantly?** Here is a table of means for a one-way ANOVA with five groups:

Group	Mean	SD	n
Group 1	146.4	22.3	30
Group 2	152.1	20.8	30
Group 3	155.8	19.8	30
Group 4	123.5	22.3	30
Group 5	127.0	22.5	30

According to the Bonferroni multiple comparisons procedure with $\alpha = 0.05$, the means for the following pairs of groups do not differ significantly: 1 and 2, 1 and 3, 2 and 3, 4 and 5. Mark the means of each pair of groups that do *not* differ significantly with the same letter. Summarize the results.

14.29 **The groups can overlap.** Refer to the previous exercise. Here is a similar table of means:

Group	Mean	SD	n
Group 1	150.2	19.1	30
Group 2	121.9	18.3	30
Group 3	129.2	18.4	30
Group 4	140.8	22.1	30
Group 5	117.2	20.6	30

According to the Bonferroni multiple comparisons procedure with $\alpha = 0.05$, the means for the following pairs of groups do not differ significantly: 1 and 4, 2 and 3, 2 and 5, 3 and 4, 3 and 5. Mark the means of each pair of groups that do *not* differ significantly with the same letter. Summarize the results.

Simultaneous confidence intervals

simultaneous confidence intervals

One way to deal with these difficulties of interpretation is to give confidence intervals for the differences. The intervals remind us that the differences are not known exactly. We want to give **simultaneous confidence intervals,** that is, intervals for all the differences among the population means with confidence (say) 95% that *all the intervals at once* cover the true population differences. Again, there are many competing procedures—in this case, many methods of obtaining simultaneous intervals.

SIMULTANEOUS CONFIDENCE INTERVALS FOR DIFFERENCES BETWEEN MEANS

Simultaneous confidence intervals for all differences $\mu_i - \mu_j$ between population means have the form

$$(\overline{x}_i - \overline{x}_j) \pm t^{**} s_p \sqrt{\frac{1}{n_i} + \frac{1}{n_j}}$$

The critical values t^{**} are the same as those used for the multiple comparisons procedure chosen.

The confidence intervals generated by a particular choice of t^{**} are closely related to the multiple comparisons results for that same method. If one of the confidence intervals includes the value 0, then that pair of means will not be declared significantly different, and vice versa.

EXAMPLE 14.25 **Software output for confidence intervals**

CASE 14.2

For simultaneous 95% Bonferroni confidence intervals, SPSS gives the output in Figure 14.16 for the data in Case 14.2. We are 95% confident that *all three* intervals simultaneously contain the true values of the population mean differences. After rounding the output, the confidence interval for the difference between the mean of the Basal group and the mean of the DRTA group is $(-10.4, -1.0)$. This interval does not include zero, so we conclude that the DRTA method results in higher mean comprehension scores than the Basal method. This is the same conclusion that we obtained from the significance test, but the confidence interval provides us with additional information about the size of the difference.

APPLY YOUR KNOWLEDGE

14.30 **Confidence interval for Basal versus Strat.** Refer to the output in Figure 14.16. Give the Bonferroni 95% confidence interval for the difference between the mean comprehension score for the Basal method and the mean comprehension score for the Strat method. Be sure to round the numbers from the output in an appropriate way. Does the interval include 0?

14.31 **Confidence interval for DRTA versus Strat.** Refer to the previous exercise. Give the interval for comparing DRTA with Strat. Does the interval include 0?

Section 14.2 Summary

- The ANOVA F test does not say which of the group means differ. It is therefore usual to add comparisons among the means to basic ANOVA.

- Specific questions formulated before examination of the data can be expressed as **contrasts.** Tests and confidence intervals for contrasts provide answers to these questions.

■ If no specific questions are formulated before examination of the data and the null hypothesis of equality of population means is rejected, **multiple comparisons** methods are used to assess the statistical significance of the differences between pairs of means. These methods are less powerful than contrasts, so use contrasts whenever a study is designed to answer specific questions.

14.3 The Power of the ANOVA Test*

The power of a test is the probability of rejecting H_0 when H_a is in fact true. Power measures how likely a test is to detect a specific alternative. When planning a study in which ANOVA will be used for the analysis, it is important to perform power calculations to check that the sample sizes are adequate to detect differences among means that are judged to be important. Power calculations also help evaluate and interpret the results of studies in which H_0 was not rejected. We sometimes find that the power of the test was so low against reasonable alternatives that there was little chance of obtaining a significant F.

In Chapter 7 we found the power for the two-sample t test. One-way ANOVA is a generalization of the two-sample t test, so it is not surprising that the procedure for calculating power is quite similar. Here are the steps that are needed:

1. Specify

   (a) an alternative (H_a) that you consider important; that is, values for the true population means $\mu_1, \mu_2, \ldots, \mu_I$;

   (b) sample sizes $n_1, n_2, \ldots, n_I$; in a preliminary study, these are usually all set equal to a common value n;

   (c) a level of significance α, usually equal to 0.05; and

   (d) a guess at the standard deviation σ.

2. Find the degrees of freedom DFG $= I - 1$ and DFE $= N - I$ and the critical value that will lead to rejection of H_0. This value, which we denote by F^*, is the upper α critical value for the $F(\text{DFG}, \text{DFE})$ distribution.

noncentrality parameter

3. Calculate the **noncentrality parameter**[5]

$$\lambda = \frac{\sum n_i(\mu_i - \overline{\mu})^2}{\sigma^2}$$

   where $\overline{\mu}$ is a weighted average of the group means,

$$\overline{\mu} = \sum w_i \mu_i$$

*This section is optional.

and the weights are proportional to the sample sizes,

$$w_i = \frac{n_i}{\sum n_i} = \frac{n_i}{N}$$

4. Find the power, which is the probability of rejecting H_0 when the alternative hypothesis is true, that is, the probability that the observed F is greater than F^*. Under H_a, the F statistic has a distribution known as the **Noncentral F distribution.** This requires special software. SAS, for example, has a function for the Noncentral F distribution. Using this function, the power is

Noncentral F distribution

$$\text{Power} = 1 - \text{PROBF}(F^*, \text{DFG}, \text{DFE}, \lambda)$$

The noncentrality parameter λ measures how far apart the means μ_i are. If the n_i are all equal to a common value n, $\overline{\mu}$ is the ordinary average of the μ_i and

$$\lambda = \frac{n\sum(\mu_i - \overline{\mu})^2}{\sigma^2}$$

If the means are all equal (the ANOVA H_0), then $\lambda = 0$. Large λ points to an alternative far from H_0, and we expect the ANOVA F test to have high power. Software makes calculation of the power quite easy, but tables and charts are also available.

EXAMPLE 14.26 **The effect of fewer subjects**

CASE 14.2

The reading comprehension study described in Cases 14.1 and 14.2 had 22 subjects in each group. Suppose that a similar study has only 10 subjects per group. How likely is this study to detect differences in the mean responses that are similar in size to those observed in the actual study?

Based on the results of the actual study, we will calculate the power for the alternative $\mu_1 = 41$, $\mu_2 = 47$, $\mu_3 = 44$, with $\sigma = 7$. The n_i are equal, so $\overline{\mu}$ is simply the average of the μ_i:

$$\overline{\mu} = \frac{41 + 47 + 44}{3} = 44$$

The noncentrality parameter is therefore

$$\begin{aligned}\lambda &= \frac{n\sum(\mu_i - \overline{\mu})^2}{\sigma^2} \\ &= \frac{(10)[(41-44)^2 + (47-44)^2 + (44-44)^2]}{49} \\ &= \frac{(10)(18)}{49} = 3.67\end{aligned}$$

Because there are three groups with 10 observations per group, DFG $= 2$ and DFE $= 27$. The critical value for $\alpha = 0.05$ is $F^* = 3.35$. The power is therefore

$$1 - \text{PROBF}(3.35, 2, 27, 3.67) = 0.3486$$

The chance that we reject the ANOVA H_0 at the 5% significance level is only about 35%.

If the assumed values of the μ_i in this example describe differences among the groups that the experimenter wants to detect, then we would want to use more than 10 subjects per group. Although H_0 is false for these μ_i, the chance of rejecting it at the 5% level is only about 35%. This chance can be increased to acceptable levels by increasing the sample sizes.

EXAMPLE 14.27 **Choosing the sample size for a future study**

CASE 14.2

To decide on an appropriate sample size for the experiment described in the previous example, we repeat the power calculation for different values of n, the number of subjects in each group. Here are the results:

n	DFG	DFE	F^*	λ	Power
20	2	57	3.16	7.35	0.65
30	2	87	3.10	11.02	0.84
40	2	117	3.07	14.69	0.93
50	2	147	3.06	18.37	0.97
100	2	297	3.03	36.73	≈ 1

With $n = 40$ the experimenters have a 93% chance of rejecting H_0 with $\alpha = 0.05$ and thereby demonstrating that the groups have different means. In the long run, 93 out of every 100 such experiments would reject H_0 at the $\alpha = 0.05$ level of significance. Using 50 subjects per group increases the chance of finding significance to 97%. With 100 subjects per group, the experimenters are virtually certain to reject H_0. The exact power for $n = 100$ is 0.99989. In most real-life situations the additional cost of increasing the sample size from 50 to 100 subjects per group would not be justified by the relatively small increase in the chance of obtaining statistically significant results.

APPLY YOUR KNOWLEDGE

14.32 **Power calculations for planning a study.** You are planning a study of the SAT mathematics scores of four groups of students. From Example 14.3, we know that the standard deviations for the three groups considered in that study were 86, 67, and 83. In Example 14.5, we found the pooled standard error to be 82.5. Since the power of the F test decreases as the standard deviation increases, use $\sigma = 86$ for the calculations in this exercise. This choice will lead to sample sizes that are perhaps a little larger than we need but will prevent us from choosing sample sizes that are too small to detect the effects of interest. You would like to conclude that the population means are different when $\mu_1 = 620$, $\mu_2 = 600$, $\mu_3 = 580$, and $\mu_4 = 560$.

(a) Pick several values for n (the number of students that you will select from each group) and calculate the power of the ANOVA F test for each of your choices.

(b) Plot the power versus the sample size. Describe the general shape of the plot.

(c) What choice of n would you choose for your study? Give reasons for your answer.

14.33 **Power against a different alternative.** Refer to the previous exercise. Repeat all parts for the alternative $\mu_1 = 610$, $\mu_2 = 600$, $\mu_3 = 590$, and $\mu_4 = 580$.

SECTION 14.3 SUMMARY

- The **power** of the F test depends upon the sample sizes, the variation among population means, and the within-group standard deviations. Some software allows easy calculation of power.

STATISTICS IN SUMMARY

Advanced statistical inference often concerns relationships among several parameters. This chapter introduces the ANOVA F test for one such relationship: equality of the means of any number of populations. The alternative to this hypothesis is "many-sided," because it allows any relationship other than "all equal." The ANOVA F test is an overall test that tells us whether the data give good reason to reject the hypothesis that all the population means are equal. Contrasts are used to answer specific questions about group means that can be formulated before the data are collected. When this is not possible, we use a multiple comparisons procedure to determine which groups differ significantly whenever the overall F test rejects the null hypothesis that all groups have the same mean. You should always accompany the ANOVA by data analysis to see what kind of inequality is present. Plotting the data in all groups side by side is particularly helpful. After studying this chapter, you should be able to do the following.

A. RECOGNITION

1. Recognize when testing the equality of several means is helpful in understanding data.
2. Recognize that the statistical significance of differences among sample means depends on the sizes of the samples and on how much variation there is within the samples.
3. Recognize when you can safely use ANOVA to compare means. Check the data production, the presence of outliers, and the sample standard deviations for the groups you want to compare.
4. Recognize when to use contrasts and multiple comparisons.
5. Recognize when it is appropriate to compute power for a one-way ANOVA.

B. INTERPRETING ANOVA

1. Explain what null hypothesis F tests in a specific setting.

2. Locate the F statistic and its P-value on the output of a computer ANOVA program.
3. Find the degrees of freedom for the F statistic from the number and sizes of the samples. Use Table E of F critical values to approximate the P-value when software does not give it.
4. If the F test is significant, use graphs and descriptive statistics to see what differences among the means are most important.
5. Formulate contrasts and analyze them using software.
6. Use multiple comparisons to determine pairs of group means that differ significantly.

CHAPTER 14 REVIEW EXERCISES

14.34 **The ANOVA framework.** In each of the following situations, identify the response variable and the populations to be compared and give I, the n_i, and N.

(a) A company wants to compare three different training programs for its new employees. In a one-month period there are 90 new hires. One-third of these are randomly assigned to each of the three training programs. At the end of the program the employees are asked to rate the effectiveness of the program on a 7-point scale.

(b) A marketing experiment compares six different types of packaging for computer disks. Each package is shown to 50 different potential consumers, who rate the attractiveness of the product on a 1 to 10 scale.

(c) Four different formulations for a new hand lotion have been produced by your product development group. You must decide which of these, if any, to market. You randomly select 400 users of your current lotion and send each new lotion to 100 of them. You ask each customer to compare the new lotion sent to them with the regular product by rating it on a 7-point scale. The middle point of the scale corresponds to no preference, while higher values indicate that the new product is preferred and lower values indicate that the regular product is better.

14.35 **The ANOVA framework.** In each of the following settings, identify the response variable and the populations to be compared and give I, the n_i, and N.

(a) A company wants to compare three different water treatment devices that can be attached to a kitchen faucet. From a list of potential customers, they select 225 households who will receive free samples. One-third of the households will receive each of the devices. Each household is asked to rate the likelihood that they would buy this kind of device on a 5-point scale.

(b) The strength of concrete depends upon the formula used to prepare it. A study compares five different mixtures. Six batches of each mixture are prepared, and the strength of the concrete made from each batch is measured.

(c) Which of three methods of teaching statistics is most effective? Twenty students are randomly assigned to each of the methods, and their scores on a final exam are recorded.

14.36 ANOVA details. For each situation in Exercise 14.34, give the following:

(a) Degrees of freedom for groups, error, and total.

(b) Null and alternative hypotheses.

(c) Numerator and denominator degrees of freedom for the F statistic.

14.37 ANOVA details. For each setting in Exercise 14.35, give the following:

(a) Degrees of freedom for groups, error, and total.

(b) Null and alternative hypotheses.

(c) Numerator and denominator degrees of freedom for the F statistic.

14.38 Pool the variances. An experiment compares three groups. The sample sizes are 10, 12, and 14, and the corresponding estimated standard deviations are 18, 24, and 20.

(a) Is it reasonable to use the assumption of equal standard deviations when we analyze these data?

(b) Give the values of the variances for the three groups and find the pooled variance s_p^2.

(c) What is the value of the pooled standard error?

14.39 Pool the variances. An experiment compares four groups. The sample sizes are 20, 220, 18, and 15 and the corresponding estimated standard deviations are 62, 40, 52, and 48.

(a) Is it reasonable to use the assumption of equal standard deviations when we analyze these data?

(b) Give the values of the variances for the four groups and find the pooled variance s_p^2.

(c) What is the value of the pooled standard error?

(d) Explain why your answer in part (c) is much closer to the standard deviation for the second group than to any of the other standard deviations.

14.40 Degrees of freedom and P-value. For each of the following situations find the degrees of freedom for the F statistic and then use Table E to approximate the P-value or use computer software to obtain an exact value.

(a) Three groups are being compared, with 8 observations per group. The value of the F statistic is 5.82.

(b) Six groups are being compared, with 11 observations per group. The value of the F statistic is 2.16.

14.41 Degrees of freedom and P-value. For each of the following situations find the degrees of freedom for the F statistic and then use Table E to approximate the P-value or use computer software to obtain an exact value.

(a) Five groups are being compared, with 13 observations per group. The value of the F statistic is 1.61.

(b) Ten groups are being compared, with 4 observations per group. The value of the F statistic is 4.68.

14.42 ANOVA in outline. Return to the change-of-majors study described in Example 14.3 (page 14-8).

(a) State H_0 and H_a for ANOVA.

(b) Outline the ANOVA table, giving the sources of variation and the degrees of freedom.

(c) What is the distribution of the F statistic under the assumption that H_0 is true?

(d) Using Table E, find the critical value for an $\alpha = 0.05$ test.

14.43 ANOVA in outline. Return to the survey of college students described in Example 14.4 (page 14-12).

(a) State H_0 and H_a for ANOVA.

(b) Outline the ANOVA table, giving the sources of variation and the degrees of freedom.

(c) What is the distribution of the F statistic under the assumption that H_0 is true?

(d) Using Table E, find the critical value for an $\alpha = 0.05$ test.

14.44 Does a product lose value when stored? Does bread lose its vitamins when stored? Small loaves of bread were prepared with flour that was fortified with a fixed amount of vitamins. After baking, the vitamin C content of two loaves was measured. Another two loaves were baked at the same time, stored for one day, and then the vitamin C content was measured. In a similar manner, two loaves were stored for three, five, and seven days before measurements were taken. The units are milligrams per hundred grams of flour (mg/100 g).[6] Here are the data:

Condition	Vitamin C (mg/100 g)	
Immediately after baking	47.62	49.79
One day after baking	40.45	43.46
Three days after baking	21.25	22.34
Five days after baking	13.18	11.65
Seven days after baking	8.51	8.13

(a) Give a table of the sample sizes, means, and standard deviations for the five conditions.

(b) Perform a one-way ANOVA for these data. State hypotheses and give the test statistic, its degrees of freedom, and the P-value.

(c) Summarize the data and the means with a plot. Use the plot and the ANOVA results to write a short summary of your conclusions.

14.45 Compare the means. Refer to the previous exercise. Use the Bonferroni or another multiple comparisons procedure to compare the group means. Summarize the results.

14.46 Two other vitamins in the product. Refer to Exercise 14.44. Measurements of the amounts of vitamin A (β-carotene) and vitamin E in each loaf are given below. Use an analysis of variance to study the data for each of these vitamins.

Condition	Vitamin A (mg/100 g)		Vitamin E (mg/100 g)	
Immediately after baking	3.36	3.34	94.6	96.0
One day after baking	3.28	3.20	95.7	93.2
Three days after baking	3.26	3.16	97.4	94.3
Five days after baking	3.25	3.36	95.0	97.7
Seven days after baking	3.01	2.92	92.3	95.1

14.47 Multiple comparisons. Refer to the previous exercise.

(a) Explain why it is inappropriate to perform a multiple comparisons analysis for the vitamin E data.

(b) Perform the Bonferroni or another multiple comparisons procedure for the vitamin A data and summarize the results.

14.48 Write a report. In Exercises 14.44 to 14.47 you have studied vitamin loss in bread stored after baking. Write a report summarizing the overall findings. Include appropriate statistical inference results and graphs.

14.49 Promotions and the expected price of a product. If a supermarket product is frequently offered at a reduced price, do customers expect the price of the product to be lower in the future? This question was examined by researchers in a study conducted on students enrolled in an introductory management course at a large midwestern university. For 10 weeks, 160 subjects read weekly ads for the same product. Students were randomly assigned to read 1, 3, 5, or 7 ads featuring price promotions during the 10-week period. They were then asked to estimate what the product's price would be the following week.[7] Table 14.3 gives the data.

(a) Make a Normal quantile plot for the data in each of the four treatment groups. Summarize the information in the plots and draw a conclusion regarding the Normality of these data.

(b) Summarize the data with a table containing the sample size, mean, and standard deviation for each group.

(c) Is the assumption of equal standard deviations reasonable here? Explain why or why not.

TABLE 14.3 Price promotion data

Number of promotions	Expected price (dollars)									
1	3.78	3.82	4.18	4.46	4.31	4.56	4.36	4.54	3.89	4.13
	3.97	4.38	3.98	3.91	4.34	4.24	4.22	4.32	3.96	4.73
	3.62	4.27	4.79	4.58	4.46	4.18	4.40	4.36	4.37	4.23
	4.06	3.86	4.26	4.33	4.10	3.94	3.97	4.60	4.50	4.00
3	4.12	3.91	3.96	4.22	3.88	4.14	4.17	4.07	4.16	4.12
	3.84	4.01	4.42	4.01	3.84	3.95	4.26	3.95	4.30	4.33
	4.17	3.97	4.32	3.87	3.91	4.21	3.86	4.14	3.93	4.08
	4.07	4.08	3.95	3.92	4.36	4.05	3.96	4.29	3.60	4.11
5	3.32	3.86	4.15	3.65	3.71	3.78	3.93	3.73	3.71	4.10
	3.69	3.83	3.58	4.08	3.99	3.72	4.41	4.12	3.73	3.56
	3.25	3.76	3.56	3.48	3.47	3.58	3.76	3.57	3.87	3.92
	3.39	3.54	3.86	3.77	4.37	3.77	3.81	3.71	3.58	3.69
7	3.45	3.64	3.37	3.27	3.58	4.01	3.67	3.74	3.50	3.60
	3.97	3.57	3.50	3.81	3.55	3.08	3.78	3.86	3.29	3.77
	3.25	3.07	3.21	3.55	3.23	2.97	3.86	3.14	3.43	3.84
	3.65	3.45	3.73	3.12	3.82	3.70	3.46	3.73	3.79	3.94

(d) Carry out a one-way ANOVA. Give the hypotheses, the test statistic with its degrees of freedom, and the P-value. Summarize your conclusion.

14.50 Compare the means. Refer to the previous exercise. Use the Bonferroni or another multiple comparisons procedure to compare the group means. Summarize the results and support your conclusions with a graph of the means.

14.51 Piano lessons and spatial-temporal reasoning. Do piano lessons improve the spatial-temporal reasoning of preschool children? In Exercise 7.70 (page 482) we examined this question by comparing the change scores (after treatment minus before treatment) of 34 children who took piano lessons with the scores of 44 children who did not. The latter group actually contained three groups of children: 10 were given singing lessons, 20 had some computer instruction, and 14 received no extra lessons. The data appear in Table 14.4.

(a) Make a table giving the sample size, mean, and standard deviation for each group.

(b) Analyze the data using one-way analysis of variance. State the null and alternative hypotheses, the test statistic with degrees of freedom, the P-value, and your conclusion.

14.52 Compare the means. Refer to the previous exercise. Use the Bonferroni or another multiple comparisons procedure to compare the group means. Summarize the results and support your conclusions with a graph of the means.

14.53 A contrast. The researchers in Exercise 14.51 based their research on a biological argument for a causal link between music and spatial-temporal reasoning. It is therefore natural to test the contrast that compares the mean of the piano lesson group with the average of the three other means. Construct this contrast, perform the significance test, and summarize the results. Note that this is not the test that we performed in Exercise 7.70 (page 482), where we did not differentiate among the three groups who did not receive piano instruction.

14.54 Exercise and healthy bones. Many studies have suggested that there is a link between exercise and healthy bones. Exercise stresses the bones and this

TABLE 14.4 Piano lesson data

Lessons	Scores									
Piano	2	5	7	−2	2	7	4	1	0	7
	3	4	3	4	9	4	5	2	9	6
	0	3	6	−1	3	4	6	7	−2	7
	−3	3	4	4						
Singing	1	−1	0	1	−4	0	0	1	0	−1
Computer	0	1	1	−3	−2	4	−1	2	4	2
	2	2	−3	−3	0	2	0	−1	3	−1
None	5	−1	7	0	4	0	2	1	−6	0
	2	−1	0	−2						

causes them to get stronger. One study examined the effect of jumping on the bone density of growing rats.[8] There were three treatments: a control with no jumping, a low-jump condition (the jump height was 30 centimeters), and a high-jump condition (60 centimeters). After 8 weeks of 10 jumps per day, 5 days per week, the bone density of the rats (expressed in mg/cm^3) was measured. Here are the data:

Group	Bone density (mg/cm^3)									
Control	611	621	614	593	593	653	600	554	603	569
Low jump	635	605	638	594	599	632	631	588	607	596
High jump	650	622	626	626	631	622	643	674	643	650

(a) Make a table giving the sample size, mean, and standard deviation for each group of rats. Is it reasonable to pool the variances?

(b) Carry out an analysis of variance. Report the F statistic with its degrees of freedom and P-value. What do you conclude?

14.55 Residuals and multiple comparisons. Refer to the previous exercise.

(a) Examine the residuals. Is the Normality assumption reasonable for these data?

(b) Use the Bonferroni or another multiple comparisons procedure to determine which pairs of means differ significantly. Summarize your results in a short report. Be sure to include a graph.

14.56 Cooking pots and dietary iron. Iron-deficiency anemia is the most common form of malnutrition in developing countries, affecting about 50% of children and women and 25% of men. Iron pots for cooking foods had traditionally been used in many of these countries, but they have been largely replaced by aluminum pots, which are cheaper and lighter. Some research has suggested that food cooked in iron pots will contain more iron than food cooked in other types of pots. One study designed to investigate this issue compared the iron content of some Ethiopian foods cooked in aluminum, clay, and iron pots.[9] One of the foods was *yesiga wet'*, beef cut into small pieces and prepared with several Ethiopian spices. The iron content of four samples of *yesiga wet'* cooked in each of the three types of pots is given below. The units are milligrams of iron per 100 grams of cooked food.

Type of pot	Iron (mg/100 g food)			
Aluminum	1.77	2.36	1.96	2.14
Clay	2.27	1.28	2.48	2.68
Iron	5.27	5.17	4.06	4.22

(a) Make a table giving the sample size, mean, and standard deviation for each type of pot. Is it reasonable to pool the variances? Note that with the small sample sizes in this experiment, we expect a large amount of variability in the sample standard deviations. For this reason, we will proceed with the ANOVA.

(b) Carry out the analysis of variance. Report the F statistic with its degrees of freedom and P-value. What do you conclude?

14.57 Residuals and multiple comparisons. Refer to the previous exercise.

(a) Examine the residuals. Is the Normality assumption reasonable for these data?

(b) Use the Bonferroni or another multiple comparisons procedure to determine which pairs of means differ significantly. Summarize your results in a short report. Be sure to include a graph.

14.58 A new material to repair wounds. One way to repair serious wounds is to insert some material as a scaffold for the body's repair cells to use as a template for new tissue. Scaffolds made from extracellular material (ECM) are particularly promising for this purpose. Because ECMs are biological material, they serve as an effective scaffold and are then absorbed. Unlike biological material that includes cells, ECMs do not trigger tissue rejection reactions in the body. One study compared 6 types of scaffold material.[10] Three of these were ECMs and the other three were made of inert materials. The subjects were three mice per scaffold type. The response measure was the percent of glucose phosphated isomerase (Gpi) cells in the region of the wound. A large value is good, indicating the presence of many bone marrow cells sent by the body to repair the tissue. Here are the data:

Material	Gpi (%)		
ECM1	55	70	70
ECM2	60	65	65
ECM3	75	70	75
MAT1	20	25	25
MAT2	5	10	5
MAT3	10	15	10

(a) Make a table giving the sample size, mean, and standard deviation for each of the six types of material. Is it reasonable to pool the variances? Note that the sample sizes are small and the data are rounded.

(b) Run the analysis of variance. Report the F statistic with its degrees of freedom and P-value. What do you conclude?

14.59 Residuals and multiple comparisons. Refer to the previous exercise.

(a) Examine the residuals. Is the Normality assumption reasonable for these data?

(b) Use the Bonferroni or another multiple comparisons procedure to determine which pairs of means differ significantly. Summarize your results in a short report. Be sure to include a graph.

14.60 A contrast. Refer to the previous two exercises. Use a contrast to compare the three ECM materials with the three other materials. Summarize your conclusions. How do these results compare with those that you obtained from the multiple comparisons procedure in the previous exercise?

14.61 What colors attract beetles? The presence of harmful insects in farm fields is detected by erecting boards covered with a sticky material and then examining the insects trapped on the board. To investigate which colors are most attractive to cereal leaf beetles, researchers placed six boards of each of four colors in a field of oats in July.[11] The table below gives data on the number of cereal leaf beetles trapped:

Color	Insects trapped					
Lemon yellow	45	59	48	46	38	47
White	21	12	14	17	13	17
Green	37	32	15	25	39	41
Blue	16	11	20	21	14	7

(a) Make a table of means and standard deviations for the four colors, and plot the means.

(b) State H_0 and H_a for an ANOVA on these data, and explain in words what ANOVA tests in this setting.

(c) Using computer software, run the ANOVA. What is the value of s_p? What are the F statistic and its P-value? What do you conclude?

14.62 Multiple comparisons. Return to the previous exercise. For the Bonferroni procedure with $\alpha = 0.05$, the value of t^{**} is 2.61. Use this multiple comparisons procedure to decide which pairs of colors are significantly different. Summarize your results. Which color would you recommend for attracting cereal leaf beetles?

14.63 Exercise and fitness. A study of the effects of exercise on physiological and psychological variables compared four groups of male subjects. The treatment group (T) consisted of 10 participants in an exercise program. A control group (C) of 5 subjects volunteered for the program but were unable to attend for various reasons. Subjects in the other two groups were selected to be similar to those in the first two groups in age and other characteristics. These were 11 joggers (J) and 10 sedentary people (S) who did not regularly exercise.[12] One of the variables measured at the end of the program was a physical fitness score. Here is part of the ANOVA table for these data:

Source	Degrees of freedom	Sum of squares	Mean square	F
Groups	3	104,855.87		
Error	32	70,500.59		
Total				

(a) Fill in the missing entries in the ANOVA table.

(b) State H_0 and H_a for this experiment.

(c) What is the distribution of the F statistic under the assumption that H_0 is true? Using Table E, give an approximate P-value for the ANOVA test. Write a brief conclusion.

(d) What is s_p^2, the estimate of the within-group variance? What is the pooled standard error s_p?

14.64 Exercise and depression. Another variable measured in the experiment described in the previous exercise was a depression score. Higher values of this score indicate more depression. Here is part of the ANOVA table for these data:

Source	Degrees of freedom	Sum of squares	Mean square	F
Groups	3		158.96	
Error	32		62.81	
Total				

(a) Fill in the missing entries in the ANOVA table.

(b) State H_0 and H_a for this experiment.

(c) What is the distribution of the F statistic under the assumption that H_0 is true? Using Table E, give an approximate P-value for the ANOVA test. What do you conclude?

(d) What is s_p^2, the estimate of the within-group variance? What is s_p?

14.65 Weight gain during pregnancy. The weight gain of women during pregnancy has an important effect on the birth weight of their children. If the weight gain is not adequate, the infant is more likely to be small and will tend to be less healthy. In a study conducted in three countries, weight gains (in kilograms) of women during the third trimester of pregnancy were measured.[13] The results are summarized in the following table:

Country	n	$\bar{x}$	s
Egypt	46	3.7	2.5
Kenya	111	3.1	1.8
Mexico	52	2.9	1.8

(a) Find the pooled estimate of the within-country variance s_p^2. What entry in an ANOVA table gives this quantity?

(b) The sum of squares for countries (groups) is 17.22. Use this information and that given above to construct an ANOVA table.

(c) State H_0 and H_a for this study.

(d) What is the distribution of the F statistic under the assumption that H_0 is true? Use Table E to find an approximate P-value for the significance test. Report your conclusion.

14.66 Food intake. In another part of the study described in the previous exercise, measurements of food intake in kilocalories were taken on many individuals several times during the period of a year. From these data, average daily food intake values were computed for each individual. The results for toddlers aged 18 to 30 months are summarized in the following table:

Country	n	$\bar{x}$	s
Egypt	88	1217	327
Kenya	91	844	184
Mexico	54	1119	285

(a) Find the pooled estimate of the within-country variance s_p^2. What entry in an ANOVA table gives this quantity?

(b) The sum of squares for countries (groups) is 6,572,551. Use this information and that given above to construct an ANOVA table.

(c) State H_0 and H_a for this study.

(d) What is the distribution of the F statistic under the assumption that H_0 is true? Use Table E to find an approximate P-value for the significance test. Report your conclusion.

14.67 **Writing contrasts.** Refer to the change-of-majors study described in Example 14.3 (page 14-8). Let μ_1, μ_2, and μ_3 represent the population mean SAT mathematics scores for the three groups.

(a) Because the first two groups (computer science, engineering and other sciences) are majoring in areas that require mathematics skills, it is natural to compare the average of these two with the third group. Write an expression in terms of the μ_i for a contrast ψ_1 that represents this comparison.

(b) Let ψ_2 be a contrast that compares the first two groups. Write an expression in terms of the μ_i for ψ_2 that represents this comparison.

14.68 **Writing contrasts.** Return to the survey of college students described in Example 14.4 (page 14-12). Let μ_1, μ_2, μ_3, and μ_4 represent the population mean expenditures on textbooks for the freshmen, sophomores, juniors, and seniors.

(a) Because juniors and seniors take higher-level courses, which might use more expensive textbooks, we want to compare the average of the freshmen and sophomores with the average of the juniors and seniors. Write a contrast that expresses this comparison.

(b) Write a contrast for comparing the freshmen with the sophomores.

(c) Write a contrast for comparing the juniors with the seniors.

14.69 **Analyzing contrasts.** Return to the SAT mathematics scores for the change-of-majors study in Example 14.3 (page 14-8). Answer the following questions for the two contrasts that you defined in Exercise 14.67.

(a) For each contrast give H_0 and an appropriate H_a. In choosing the alternatives you should use information given in the description of the problem, but you may not consider any impressions obtained by inspection of the sample means.

(b) Find the values of the corresponding sample contrasts c_1 and c_2.

(c) Using the value $s_p = 82.5$, calculate the standard errors s_{c_1} and s_{c_2} for the sample contrasts.

(d) Give the test statistics and approximate P-values for the two significance tests. What do you conclude?

(e) Compute 95% confidence intervals for the two contrasts.

14.70 **High school grades and changing majors.** The following table presents data on the high school mathematics grades of the students in the change-of-majors study described in Example 14.3 (page 14-8). The sample sizes in this exercise are different from those in Example 14.3 because complete information was not available on all students. The grades have been coded so that 10 = A, 9 = A−, and so on.

Second-year major	n	$\overline{x}$	s
Computer science	90	8.77	1.41
Engineering and other sciences	28	8.75	1.46
Other	106	7.83	1.74

The two contrasts that you defined in Exercise 14.67 naturally lead to questions about high school mathematics grades.

(a) For each contrast state H_0 and an appropriate H_a. In choosing the alternatives you should use information given in the description of the problem, but you may not consider any impressions obtained by inspection of the sample means.

(b) Find the values of the corresponding sample contrasts c_1 and c_2.

(c) Using the value $s_p = 1.581$, calculate the standard errors s_{c_1} and s_{c_2} for the sample contrasts.

(d) Give the test statistics and approximate P-values for the two significance tests. What do you conclude?

(e) Compute 95% confidence intervals for the two contrasts.

14.71 **Contrasts for exercise and fitness.** In the exercise program study described in Exercise 14.63, the summary statistics for physical fitness scores are as follows:

Group	n	$\overline{x}$	s
Treatment (T)	10	291.91	38.17
Control (C)	5	308.97	32.07
Joggers (J)	11	366.87	41.19
Sedentary (S)	10	226.07	63.53

The researchers wanted to address the following questions for the physical fitness scores. In these questions "better" means a higher fitness score. (1) Is T better than C? (2) Is T better than the average of C and S? (3) Is J better than the average of the other three groups?

(a) For each of the three questions, define an appropriate contrast. Translate the questions into null and alternative hypotheses about these contrasts.

(b) Test your hypotheses and give approximate P-values. Summarize your conclusions. Do you think that the use of contrasts in this way gives an adequate summary of the results?

(c) You found that the groups differ significantly in the physical fitness scores. Does this study allow conclusions about causation—for example, that a sedentary lifestyle causes people to be less physically fit? Explain your answer.

14.72 Contrasts for exercise and depression. Exercise 14.64 gives the ANOVA table for depression scores from the exercise program study described in Exercise 14.63. Here are the summary statistics for the depression scores:

Group	n	$\bar{x}$	s
Treatment (T)	10	51.90	6.42
Control (C)	5	57.40	10.46
Joggers (J)	11	49.73	6.27
Sedentary (S)	10	58.20	9.49

In planning the experiment, the researchers wanted to address the following questions for the depression scores. In these questions "better" means a lower depression score. (1) Is T better than C? (2) Is T better than the average of C and S? (3) Is J better than the average of the other three groups?

(a) For each of the three questions, define an appropriate contrast. Translate the questions into null and alternative hypotheses about these contrasts.

(b) Test your hypotheses and give approximate P-values. Summarize your conclusions. Do you think that the use of contrasts in this way gives an adequate summary of the results?

(c) You found that the groups differ significantly in the depression scores. Does this study allow conclusions about causation—for example, that a sedentary lifestyle causes people to be more depressed? Explain your answer.

14.73 Multiple comparisons for exercise and fitness. Refer to the physical fitness scores for the four groups in the exercise program study discussed in Exercises 14.63 and 14.71. Computer software gives the critical value for the Bonferroni multiple comparisons procedure with $\alpha = 0.05$ as $t^{**} = 2.81$. Use this procedure to compare the mean fitness scores for the four groups. Summarize your conclusions.

14.74 Multiple comparisons for exercise and depression. Refer to the depression scores for the four groups in the exercise program study discussed in Exercises 14.64 and 14.72. Computer software gives the critical value for the Bonferroni multiple comparisons procedure with $\alpha = 0.05$ as $t^{**} = 2.81$. Use this procedure to compare the mean depression scores for the four groups. Summarize your conclusions.

14.75 (Optional) Power for weight gain during pregnancy. You are planning a study of the weight gains of women during the third trimester of pregnancy similar to that described in Exercise 14.65. The standard deviations given in that exercise range from 1.8 to 2.5. To perform power calculations, assume that the standard deviation is $\sigma = 2.4$. You have three groups, each with n

subjects, and you would like to reject the ANOVA H_0 when the alternative $\mu_1 = 2.6$, $\mu_2 = 3.0$, and $\mu_3 = 3.4$ is true. Use software to make a table of powers against this alternative (similar to the table in Example 14.27, page 14-42) for the following numbers of women in each group: $n = 50, 100, 150, 175$, and 200. What sample size would you choose for your study?

14.76 **(Optional) More power.** Repeat the previous exercise for the alternative $\mu_1 = 2.7$, $\mu_2 = 3.1$, and $\mu_3 = 3.5$. Why are the results the same?

14.77 **Transform amounts to percents.** Refer to Exercise 14.44 (page 14-46), where we studied the effects of storage on the vitamin C content of bread. In this experiment 64 mg/100 g of vitamin C was added to the flour that was used to make each loaf.

(a) Convert the vitamin C amounts (mg/100 g) to percents of the amounts originally in the loaves by dividing the amounts in Exercise 14.44 by 64 and multiplying by 100. Calculate the transformed means, standard deviations, and standard errors and summarize them with the sample sizes in a table.

(b) Explain how you could have calculated the table entries directly from the table you gave in part (a) of Exercise 14.44.

(c) Analyze the percents using analysis of variance. Compare the test statistic, degrees of freedom, P-value, and conclusion you obtain here with the corresponding values that you found in Exercise 14.44.

14.78 **Changing units and ANOVA.** Refer to the previous exercise. Suggest a general conclusion about what happens to the F test statistic, degrees of freedom, P-value, and conclusion when you perform analysis of variance on data after changing the units of measurement. (A change of units is a linear transformation that multiplies each observation by the same positive constant and perhaps adds another constant.)

14.79 **The effect of an outlier.** Refer to the color attractiveness experiment described in Exercise 14.61 (page 14-50). Suppose that when entering the data into the computer, you accidentally entered the first observation as 450 rather than 45.

(a) Run the ANOVA with the incorrect observation. Summarize the results.

(b) Compare this run with the results obtained with the correct data set. What does this illustrate about the effect of outliers in an ANOVA?

(c) Compute a table of means and standard deviations for each of the four treatments using the incorrect data. How does this table help you to detect the incorrect observation?

14.80 **Try a transformation.** Refer to the color attractiveness experiment described in Exercise 14.61 (page 14-50). The square root transformation is often used for variables that are counts, such as the number of insects trapped in this example. In many cases data transformed in this way will conform more closely to the assumptions of Normality and equal standard deviations.

(a) Make a table of the new data after taking square roots. Find the group means and standard deviations. Are the standard deviations of the transformed data more nearly equal than those of the original data?

(b) Carry out the ANOVA for the transformed data. Compare your conclusions with the original ANOVA in Exercise 14.61.

14.81 Regression or ANOVA? Refer to the price promotion study that we examined in Exercise 14.49 (page 14-47). The explanatory variable in this study is the number of price promotions in a 10-week period, with possible values of 1, 3, 5, and 7. ANOVA treats the explanatory variable as categorical—it just labels the groups to be compared. In this study the explanatory variable is in fact quantitative, so we could use simple linear regression rather than one-way ANOVA if there is a linear pattern.

(a) Make a scatterplot of the responses against the explanatory variable. Is the pattern roughly linear?

(b) In ANOVA, the *F* test null hypothesis states that groups have no effect on the mean response. What test in regression tests the null hypothesis that the explanatory variable has no linear relationship with the response?

(c) Carry out the regression. Compare your results with those from the ANOVA in Exercise 14.49. Are there any reasons—perhaps from part (a)—to prefer one or the other analysis?

CHAPTER 14 CASE STUDY EXERCISES

CASE STUDY 14.1: **Compare three products for treating dandruff.** Analysis of variance methods are often used to compare different products. In medical settings, studies to assess the effectiveness of one or more treatments for a particular medical condition are often called *clinical trials*. A clinical trial compared the effectiveness of three products and a placebo in reducing dandruff. The products were 1% pyrithione zinc shampoo (PyrI), the same shampoo but with instructions to shampoo two times (PyrII), 2% ketoconazole shampoo (Keto), and a placebo shampoo (Placebo). After six weeks of treatment, eight sections of the scalp were examined and given a score that measured the amount of scalp flaking on a 0 to 10 scale. The response variable was the sum of these eight scores.

An analysis of the baseline flaking measure indicated that randomization of patients to treatments was successful in that no differences were found among the groups. At baseline there were 112 subjects in each of the three treatment groups and 28 subjects in the Placebo group. During the clinical trial, 3 dropped out from the PyrII group and 6 from the Keto group. No patients dropped out of the other two groups. The data are given in the DANDRUFF data set described in the Data Appendix. Analyze the data using the methods you learned in this chapter. Be sure to include numerical and graphical summaries, a detailed explanation of the statistical model used, and a summary of all results. Include an analysis of the residuals to examine the assumptions of your model. Include an analysis of the following contrasts: (1) Placebo versus the average of the three treatments; (2) Keto versus the average of the two Pyr treatments; and (3) PyrI versus PyrII.

CASE STUDY 14.2: **Compare the three educational products.** Cases 14.1 (page 14-15) and 14.2 (page 14-27) compare pretest and posttest reading comprehension scores for three different products designed to teach reading to young children. In the actual study there were 2 pretest measures and 3 posttest measures. The data are given in the READING data set described in the Data Appendix. The pretest variable analyzed in Case 14.1 is Pre1 and the posttest variable analyzed in Case 14.2 is Post3. Analyze the other pretest variable and the other 2 posttest variables. Use the exercises in the text to guide your analyses. Summarize the results for all 5 variables in a report. In particular, do the pretests show that the three groups of subjects were similar at the start of the study? Do the posttests point to one or more of the teaching methods as more effective than the others?

Notes for Chapter 14

1. Results of the study are reported in P. F. Campbell and G. P. McCabe, "Predicting the success of freshmen in a computer science major," *Communications of the ACM,* 27 (1984), pp. 1108–1113.

2. This rule is intended to provide a general guideline for deciding when serious errors may result by applying ANOVA procedures. When the sample sizes in each group are very small, the sample variances will tend to vary much more than when the sample sizes are large. In this case, the rule may be a little too conservative. For unequal sample sizes, particular difficulties can arise when a relatively small sample size is associated with a population having a relatively large standard deviation. Careful judgment is needed in all cases. By considering P-values rather than fixed level α testing, judgments in ambiguous cases can more easily be made; for example, if the P-value is very small, say, 0.001, then it is probably safe to reject H_0 even if there is a fair amount of variation in the sample standard deviations.

3. This example is based on data from a study conducted by Jim Baumann and Leah Jones of the Purdue University School of Education.

4. This example is based on Amna Kirmani and Peter Wright, "Money talks: perceived advertising expense and expected product quality," *Journal of Consumer Research,* 16 (1989), pp. 344–353.

5. Several different definitions for the noncentrality parameter of the noncentral F distribution are in use. When $I = 2$, the λ defined here is equal to the square of the noncentrality parameter δ that we used for the two-sample t test in Chapter 7. Many authors prefer $\phi = \sqrt{\lambda/I}$. We have chosen to use λ because it is the form needed for the SAS function PROBF.

6. These data were provided by Helen Park; see H. Park et al., "Fortifying bread with each of three antioxidants," *Cereal Chemistry,* 74 (1997), pp. 202–206.

7. Based on M. U. Kalwani and C. K. Yim, "Consumer price and promotion expectations: an experimental study," *Journal of Marketing Research,* 29 (1992), pp. 90–100.

8. Data provided by Jo Welch of the Purdue University Department of Foods and Nutrition.

9. Based on A. A. Adish et al., "Effect of consumption of food cooked in iron pots on iron status and growth of young children: a randomised trial," *The Lancet,* 353 (1999), pp. 712–716.

10. See S. Badylak et al., "Marrow-derived cells populate scaffolds composed of xenogeneic extracellular matrix," *Experimental Hematology,* 29 (2001), pp. 1310–1318.

11. Modified from M. C. Wilson and R. E. Shade, "Relative attractiveness of various luminescent colors to the cereal leaf beetle and the meadow spittlebug," *Journal of Economic Entomology,* 60 (1967), pp. 578–580.

12. Data provided by Dennis Lobstein, from his Ph.D. dissertation, "A multivariate study of exercise training effects on beta-endorphin and emotionality in psychologically normal, medically healthy men," Purdue University, 1983.

13. These data were taken from Collaborative Research Support Program in Food Intake and Human Function, *Management Entity Final Report,* University of California, Berkeley, 1988.

Chapter 14

14.1 $I = 3$, $N = 60$, $n_1 = 20$, $n_2 = 20$, and $n_3 = 20$. The ANOVA model is $x_{ij} = \mu_i + \varepsilon_{ij}$. The ε_{ij} are assumed to be from an $N(0, \sigma)$ distribution where σ is the common standard deviation. The parameters of the model are the I population means μ_1, μ_2, and μ_3, and the common standard deviation σ.

14.3 **(a)** The largest standard deviation is 120 and the smallest is 80. The ratio of these is 120/80 = 1.5. This ratio is less than 2, so it is reasonable to pool the standard deviations for these data. **(b)** $I = 3, \bar{x}_1 = 150, \bar{x}_2 = 175, \bar{x}_3 = 200$. We estimate σ by $s_p = 101.32$

14.5 Stem plots for the three groups are given below.

Basal		DRTA		Strat	
0	4	0		0	445
0	67	0	6777	0	666777
0	888999	0	888889999	0	889
1	01	1	000	1	111
1	22222223	1	2233	1	22333
1	45	1	5	1	44
1	6	1	6	1	

The distribution for the Basal group appears to be centered at a slightly larger score than that for the DRTA group and perhaps for the Strat group. The distribution of the DRTA scores shows some right-skewness. There are no clear outliers in any of the groups.

14.7 SSG + SSE = 20.58 + 572.45 = 593.03 = SST.

14.9 DFG + DFE = 2 + 63 = 65 = DFT.

14.11 MST = 9.124. Using statistical software (we used Minitab) we find the mean of all 66 observations in Table 14.1 to be 9.788 and the variance of all 66 observations in Table 14.1 to be 9.125.

14.13 **(a)** The ANOVA F statistic has 3 numerator degrees of freedom and 20 denominator degrees of freedom. **(b)** The F statistic would need to be larger than 3.10 to have a P-value less than 0.05.

14.15 **(a)** It is always true that $s_p^2 = \text{MSE}$. Hence, $s_p = \sqrt{\text{MSE}}$. Because of this, SAS calls s_p "Root MSE." **(b)** $s_p = \sqrt{\text{MSE}}$. In Excel, MSE is found in the ANOVA table in the row labeled "Within Groups" and under the column labeled MS. Take the square root of this entry (0.616003 in this case) to get s_p.

14.17 Try the applet. You should find that the F statistic increases and the P-value decreases if the pooled standard error is kept fixed and the variation among the group means increases.

14.19 The standard error for the contrast $SE_c = 2$.

14.21 The 95% confidence interval is $15 \pm 3.968 = (11.032, 18.968)$.

14.23 $$t_{23} = \frac{46.7273 - 44.2727}{6.31\sqrt{\frac{1}{22} + \frac{1}{22}}} = 1.29.$$

14.25 The value of t^{**} for the Bonferroni procedure when $\alpha = 0.05$ is $t^{**} = 2.46$ (see Example 14.22). $t_{23} = 1.29$. Because $|1.29| < t^{**}$, using the Bonferroni procedure we would not reject the null hypothesis that the population means for groups 2 and 3 are different.

14.27 From the output in Figure 14.16 $\bar{x}_2 - \bar{x}_3 = 2.4545$ and the standard error for this difference is 1.90378. Thus, $t_{23} = 2.4545/1.90378 = 1.29$.

14.29

Group	Mean	SD	n
Group 1	150.2^{a}	19.1	30
Group 2	$121.9^{b,c}$	18.3	30
Group 3	$129.2^{b,d,e}$	18.4	30
Group 4	$140.8^{a,d}$	22.1	30
Group 5	$117.2^{c,e}$	20.6	30

We see that Groups 1 and 4 have the largest means. Group 1 differs from Groups 2, 3, and 5. Group 4 differs from Groups 2 and 5.

14.31 The Bonferroni 95% confidence interval is (−2.2, 7.1). This interval includes 0.

14.33 **(a)**

n	DFG	DFE	F^*	λ	Power
10	3	36	2.8663	0.676	0.0883
20	3	76	2.7249	1.352	0.1366
30	3	116	2.6828	2.028	0.1891
40	3	156	2.6626	2.704	0.2444
50	3	196	2.6507	3.380	0.3010
100	3	396	2.6274	6.76	0.5684
150	3	596	2.6198	10.14	0.7645
200	3	796	2.6160	13.52	0.8829
250	3	996	2.6139	16.90	0.9458
500	3	1996	2.6095	33.80	0.9994

(b) The power increases as sample size increases, but the amount of increase gradually levels off as sample size increases. **(c)** Large sample sizes are needed to obtain high power. If the cost per student is high, one might consider a sample size of 150, with a power of 0.7645. If this is not adequate, one might consider a sample size of 200, with a power of 0.8829. If it is important to have a fairly large power, one might consider a sample size of 250, with a power of 0.9458.

14.35 (**a**) The response variable is the rating (on a 5-point scale) of the likelihood the household would buy the water treatment device. There are three populations, and these are the responses of all people to each of the three water treatment devices. $I = 3$. $N = 225$. $n_1 = n_2 = n_3 = 75$. (**b**) The response variable is the strength of the concrete. There are five populations, and these are the strengths of all possible batches of concrete for each of the five mixtures. $I = 5$. $n_1 = n_2 = n_3 = n_4 = n_5 = 6$. $N = 30$. (**c**) The score on a final exam is the response variable. There are three populations, and these are the final-exam scores of all people after being taught by each of the three methods. $I = 3$. $n_1 = n_2 = n_3 = 20$. $N = 60$.

14.37 (**a**) For the first setting, DFG = 2, DFE = 222, DFT = 224. For the second setting, DFG = 4, DFE = 25, DFT = 29. For the third setting, DFG = 2, DFE = 57, DFT = 59. (**b**) For the first setting, the null and alternative hypotheses are H_0: $\mu_1 = \mu_2 = \mu_3$, H_a: not all of the μ_i are equal. For the second setting, the null and alternative hypotheses are $H_0 : \mu_1 = \mu_2 = \mu_3 = \mu_4 = \mu_5$, H_a : not all of the μ_i are equal. For the third setting, the null and alternative hypotheses are H_0: $\mu_1 = \mu_2 = \mu_3$, H_a: not all of the μ_i are equal. (**c**) For the first setting, numerator degrees of freedom for the F statistic = DFG = 2, denominator degrees of freedom for the F statistic = DFE = 222. For the second setting, numerator degrees of freedom for the F statistic = DFG = 4, denominator degrees of freedom for the F statistic = DFE = 25. For the third setting, numerator degrees of freedom for the F statistic = DFG = 2, denominator degrees of freedom for the F statistic = DFE = 57.

14.39 (**a**) The ratio of the largest to the smallest standard deviation is 62/40 = 1.55, which is less than 2. It is reasonable to assume equal standard deviations when we analyze these data. (**b**) $s_1^2 = 3844$, $s_2^2 = 1600$, $s_3^2 = 2704$, $s_4^2 = 2304$, $s_p^2 = 1864.91$. (**c**) $s_p = 43.18$. (**d**) The sample size in the second group is much larger than that for all other groups combined. Thus, the variance for this group is weighted more heavily in the formula for s_p^2, and so s_p^2 will be closer to the variance for the second group than to the variance for any other group.

14.41 (**a**) The F statistic has 4 numerator degrees of freedom and 60 denominator degrees of freedom. The approximate P-value is larger than 0.100. Using statistical software, the exact P-value is 0.1835. (**b**) The F statistic has 9 numerator degrees of freedom and 30 denominator degrees of freedom. The approximate P-value is smaller than 0.001. Using statistical software, the exact P-value is 0.0006.

14.43 (**a**) $H_0 : \mu_1 = \mu_2 = \mu_3 = \mu_4$, H_a: not all of the μ_i are equal.

(**b**)

Source	Degrees of freedom	Sum of squares	Mean square	F
Groups	DFG = 3	SSG	MSG = SSG/DFG	MSG/MSE
Error	DFE = 196	SSE	MSE = SSE/DFE	
Total	DFT = 199	SST	MST = SST/DFT	

(**c**) Under the assumption that H_0 is true, the F statistic has the $F(3, 196)$ distribution. (**d**) 2.70 is the (approximate) critical value for an $\alpha = 0.05$ test.

14.45

Bonferroni Tests

Condition $i - j$	Difference	Std. error	P-value
1 − 0	−6.75000	1.321	0.036734
3 − 0	−26.9100	1.321	0.000053
3 − 1	−20.1600	1.321	0.000219
5 − 0	−36.2900	1.321	0.000012
5 − 1	−29.5400	1.321	0.000033
5 − 3	−9.38000	1.321	0.008542
7 − 0	−40.3850	1.321	0.000007
7 − 1	−33.6350	1.321	0.000017
7 − 3	−13.4750	1.321	0.001551
7 − 5	−4.09500	1.321	0.238149

The differences in the group means are all significantly different from 0 at the $\alpha = 0.05$ level except the difference after 5 and 7 days. At the $\alpha = 0.01$ level, all the differences are significantly different from 0 except the difference between 0 and 1 days after baking, and the difference between 5 and 7 days after baking.

14.47 **(a)** The ANOVA in Exercise 14.46 did not reject the hypothesis at the 0.05 level that any of the group means differed. Thus, no further analysis on which group means differed is appropriate. **(b)**

Bonferroni Tests

Condition $i - j$	Difference	Std. error	P-value
1 − 0	−0.110000	0.0608	0.752553
3 − 0	−0.140000	0.0608	0.514112
3 − 1	−0.030000	0.0608	0.999966
5 − 0	−0.045000	0.0608	0.998871
5 − 1	0.065000	0.0608	0.982858
5 − 3	0.095000	0.0608	0.861026
7 − 0	−0.385000	0.0608	0.014421
7 − 1	−0.275000	0.0608	0.061029
7 − 3	−0.245000	0.0608	0.096013
7 − 5	−0.340000	0.0608	0.025003

The only group means that are significantly different at the $\alpha = 0.05$ level are the difference between 0 and 7 days after baking and between 7 and 5 days after baking.

14.49 **(a)** The plots show that Group 2 has a modest outlier, but otherwise there are no serious departures from Normality. The assumption that the data are (approximately) Normal is not unreasonable.

(b)

Number of promotions	Sample size	Mean	Std. dev.
1	40	4.22400	0.273410
3	40	4.06275	0.174238
5	40	3.75900	0.252645
7	40	3.54875	0.275031

(c) The ratio of the largest to the smallest is 0.275031/0.174238 = 1.58. This is less than 2, so it is not unreasonable to assume that the population standard deviations are equal. (d) The hypothesee for AVOVA are H_0: $\mu_1 = \mu_2 = \mu_3 = \mu_4$, H_a: not all of the μ_i are equal. The ANOVA table is

Source	Degrees of freedom	Sum of squares	Mean square	*F*	*P*-value
Groups	3	10.9885	3.66285	59.903	≤ 0.0001
Error	156	9.53875	0.061146		
Total	159	20.5273			

The F statistic has 3 numerator and 156 denominator degrees of freedom. The P-value is ≤ 0.0001, and we would conclude that there is strong evidence that the population mean expected prices associated with the different numbers of promotions are not all equal.

14.51 (a)

Group	Sample size	Mean	Std. dev.
Piano	34	3.61765	3.05520
Singing	10	−0.300000	1.49443
Computer	20	0.450000	2.21181
None	14	0.785714	3.19082

(b) The hypotheses for ANOVA are H_0: $\mu_1 = \mu_2 = \mu_3 = \mu_4$, H_a: not all of the μ_i are equal. The ANOVA table is

Source	Degrees of freedom	Sum of squares	Mean square	*F*	*P*-value
Groups	3	207.281	69.0938	9.2385	≤ 0.0001
Error	74	553.437	7.47887		
Total	77	760.718			

The F statistic has 3 numerator and 74 denominator degrees of freedom. The P-value is ≤ 0.0001, and we would conclude that there is strong evidence that the population mean changes in scores associated with the different types of instruction are not all equal.

14.53 The contrast is $\psi = \mu_1 - (1/3)\mu_2 - (1/3)\mu_3 - (1/3)\mu_4$. To test the null hypothesis H_0: $\psi = 0$, the t statistic is $t = c/\text{SE}_c = 3.306/0.636 = 5.20$. P-value ≤ 0.001. We conclude that there is strong statistical evidence that the mean of the piano group differs from the average of the means of the other three groups.

14.55 (a) Plots show no serious departures from Normality, so the Normality assumption is reasonable.

(b)

Bonferroni Tests

Group $i - j$	Difference	Std. error	*P*-value
2 − 1	11.4000	9.653	0.574592
3 − 1	37.6000	9.653	0.001750
3 − 2	26.2000	9.653	0.033909

At the $\alpha = 0.05$ level we see that Group 3 (the high-jump group) differs from the other two. The other two groups (the control group and the low-jump group) are not significantly different. It appears that the mean density after 8 weeks is different (higher) for the high-jump group than for the other two.

14.57 (a) Plots show no serious departures from Normality, so the Normality assumption is reasonable.

(b)

	Bonferroni Tests		
Group $i - j$	Difference	Std. error	P-value
2 − 1	0.120000	0.3751	0.985534
3 − 1	2.62250	0.3751	0.000192
3 − 2	2.50250	0.3751	0.000274

At the $\alpha = 0.05$ level we see that Group 3 (the iron pots) differs from the other two. The other two groups (the aluminum and clay pots) are not significantly different. It appears that the mean iron content of *yesiga wet'* when it is cooked in iron pots is different (higher) than when it is cooked in the other two.

14.59 (a) Plots show no serious departures from Normality (but note the granularity of the Normal probability plot. It appears that values are rounded to the nearest 5%), so the Normality assumption is not unreasonable.

(b)

	Bonferroni Tests		
Group $i - j$	Difference	Std. error	P-value
ECM2 − ECM1	−1.66667	3.600	1.00000
ECM3 − ECM1	8.33333	3.600	0.450687
ECM3 − ECM2	10.0000	3.600	0.223577
MAT1 − ECM1	−41.6667	3.600	0.000001
MAT1 − ECM2	−40.0000	3.600	0.000002
MAT1 − ECM3	−50.0000	3.600	0.000000
MAT2 − ECM1	−58.3333	3.600	0.000000
MAT2 − ECM2	−56.6667	3.600	0.000000
MAT2 − ECM3	−66.6667	3.600	0.000000
MAT2 − MAT1	−16.6667	3.600	0.008680
MAT3 − ECM1	−53.3333	3.600	0.000000
MAT3 − ECM2	−51.6667	3.600	0.000000
MAT3 − ECM3	−61.6667	3.600	0.000000
MAT3 − MAT1	−11.6667	3.600	0.101119
MAT3 − MAT2	5.00000	3.600	0.957728

At the $\alpha = 0.05$ level we see that none of the ECMs differ from each other, that all the ECMs differ from all the other types of materials (the MATs), and that MAT1 and MAT2 differ from each other. The most striking differences are those between the ECMs and the other materials.

14.61 (a)

Group	Sample size	Mean	Std. dev.
Lemon	6	47.1667	6.79461
White	6	15.6667	3.32666
Green	6	31.5000	9.91464
Blue	6	14.8333	5.34478

(b) The hypotheses are $H_0: \mu_1 = \mu_2 = \mu_3 = \mu_4$, H_a: not all of the μ_i are equal. ANOVA tests whether the mean number of insects trapped by the different colors are the same or if they differ. If they differ, ANOVA does not tell us which ones differ.

(c)

Source	Degrees of freedom	Sum of squares	Mean square	F	P-value
Groups	3	4218.46	1406.15	30.552	≤ 0.0001
Error	20	920.500	46.0250		
Total	23	5138.96			

$s_p = 6.78$. We conclude that there is strong evidence of a difference in the mean number of insects trapped by the different colors.

14.63 (a)

Source	Degrees of freedom	Sum of squares	Mean square	F
Groups	3	104,855.87	34,951.96	15.86
Error	32	70,500.59	2,203.14	
Total	35	175,356.46	5,010.18	

(b) H_0: $\mu_1 = \mu_2 = \mu_3 = \mu_4$, H_a: not all of the μ_i are equal. (c) The F statistic has the $F(3, 32)$ distribution. The P-value is smaller than 0.001. (d) $s_p^2 = 2,203.14$, $s_p = 46.94$.

14.65 (a) $s_p^2 = 3.90$. This quantity corresponds to MSE in the ANOVA table.
(b)

Source	Degrees of freedom	Sum of squares	Mean square	F
Groups	2	17.22	8.61	2.21
Error	206	803.4	3.90	
Total	208	175,356.46	5010.18	

(c) $H_0: \mu_1 = \mu_2 = \mu_3$, H_a: not all of the μ_i are equal. (d) The F statistic has the $F(2, 206)$ distribution. The P value is greater than 0.100. We conclude that these data do not provide evidence that the mean weight gains of pregnant women in these three countries differ.

14.67 (a) A contrast is $\psi_1 = (0.5)\mu_1 + (0.5)\mu_2 - \mu_3$. (b) A contrast is $\psi_2 = \mu_1 - \mu_2$.

14.69 (a) For the contrast $\psi_1 = (0.5)\mu_1 + (0.5)\mu_2 - \mu_3$, it would be reasonable to test the hypotheses $H_0: \psi_1 = 0$, $H_a: \psi_1 > 0$. For the contrast $\psi_2 = \mu_1 - \mu_2$, it would be reasonable to test the hypotheses H_0: $\psi_2 = 0$, H_a: $\psi_2 \neq 0$.

(b) Estimate of ψ_1: $c_1 = 49$. Estimate of ψ_2: $c_2 = -10$. (c) $SE_{c_1} = 6.32$, $SE_{c_2} = 7.29$. (d) The test statistic for H_0: $\psi_1 = 0$, H_a: $\psi_1 > 0$, is $t = 7.75$. P-value < 0.005. We conclude that there is strong evidence that the average of the mean SAT mathematics scores for computer science and engineering and other sciences majors is larger than the mean SAT mathematics scores for all other majors.

For H_0: $\psi_2 = 0$, H_a: $\psi_2 \neq 0$, the test statistic is $t = 1.37$. $0.10 <$ P-value < 0.20. We conclude that there is not strong evidence that the average of the mean SAT mathematics scores for computer science majors differs from that for engineering and other science majors. (e) A 95% confidence interval for ψ_1 is $49 \pm 12.54 = (36.46, 61.54)$. A 95% confidence interval for ψ_2 is $-10 \pm 14.46 = (-24.46, 4.46)$.

14.71 (a) Question 1. Contrast: $\psi_1 = \mu_1 - \mu_2$. Hypotheses: H_0: $\psi_1 = 0$, H_a: $\psi_1 > 0$. Question 2. Contrast: $\psi_2 = \mu_1 - (0.5)\mu_2 - (0.5)\mu_4$. Hypotheses: H_0: $\psi_2 = 0$, H_a: $\psi_2 > 0$. Question 3. Contrast: $\psi_3 = \mu_3 - (1/3)\mu_1 - (1/3)\mu_2 - (1/3)\mu_4$. Hypotheses: H_0: $\psi_3 = 0$, H_a: $\psi_3 > 0$. (b) Question 1: P-value > 0.25. We do not have evidence that T is better than C. Question 2: $0.10 <$ P-value < 0.15. There is at best weak evidence that T is better than the average of C and S. Question 3: P-value < 0.005. There is strong evidence that J is better than the average of the other three groups. (c) This is an observational study. Males were not assigned at random to treatments. Thus, although the researchers tried to match those in the groups with respect to age and other characteristics, there are reasons why people choose to jog or choose to be sedentary that may affect other aspects of their health. It is always risky to draw conclusions of causality from a single (small) observational study, no matter how well designed it is in other respects.

14.73 For each pair of means we get the following: Group 1 (T) vs. Group 2 (C): $t_{12} = -0.66$. $|-.66|$ is not larger than $t^{**} = 2.81$, so we do not have strong evidence that T and C differ. Group 1 (T) vs. Group 3 (J): $t_{13} = -3.65$. $|-3.65|$ is larger than $t^{**} = 2.81$, so we have strong evidence that T and J differ. Group 1 (T) vs. Group 4 (S): $t_{14} = 3.14$. $|3.14|$ is larger than $t^{**} = 2.81$, so we have strong evidence that T and S differ. Group 2 (C) vs. Group 3 (J): $t_{23} = -2.29$. $|-2.29|$ is not larger than $t^{**} = 2.81$, so we do not have strong evidence that C and J differ. Group 2 (C) vs. Group 4 (S): $t_{24} = 3.22$. $|3.22|$ is larger than $t^{**} = 2.81$, so we have strong evidence that C and S differ. Group 3 (J) vs. Group 4 (S): $t_{34} = 6.86$. $|6.86|$ is larger than $t^{**} = 2.81$, so we have strong evidence that J and S differ.

14.75

n	DFG	DFE	F^*	λ	Power
50	2	147	3.0576	2.78	0.2950
100	2	297	3.0261	5.56	0.5453
150	2	447	3.0158	8.34	0.7336
175	2	522	3.0130	9.73	0.8017
200	2	597	3.0108	11.12	0.8548

A sample size of 175 gives reasonable power. The gain in power by using 200 women per group may not be worthwhile unless it is easy to get women for the study. If it is difficult or expensive to include more women in the study, one might consider a sample size of 150 per group.

14.77 **(a)**

Group	Sample size	Mean	Std. dev.	Std. error
0	2	76.1016	2.39753	1.695
1	2	65.5547	3.32561	2.352
3	2	34.0547	1.20429	0.852
5	2	19.3984	1.69043	1.195
7	2	13	0.419845	0.297

(b) The sample sizes would be the same in both tables. The means, standard deviations, and standard errors above could have been obtained from those in Exercise 14.44 by dividing each by 64 and multiplying the result by 100. Means, standard deviations, and standard errors change the same way as individual values do. **(c)** The ANOVA table is

Source	Degrees of freedom	Sum of squares	Mean square	*F*	*P*-value
Groups	4	6263.97	1565.99	367.74	≤ 0.0001
Error	5	21.2920	4.25840		
Total	9	6285.26			

We conclude that there is strong evidence that the group means differ, that is, that the mean % vitamin C content is not the same for all conditions. The degrees of freedom, the *F* statistic, and *P*-value are all the same as in Exercise 14.44.

14.79 **(a)** The ANOVA table with the incorrect observation is

Source	Degrees of freedom	Sum of squares	Mean Square	*F*	*P*-value
Groups	3	40,820.3	13606.8	2.0032	0.1460
Error	20	135,853	6792.65		
Total	23	176,673			

The *P*-value is larger than 0.10, so we would conclude that we do not have strong evidence that the mean number of insects that will be trapped differs between the different-colored traps. **(b)** The results are very different. In Exercise 14.61, *P*-value ≤ 0.0001, and we concluded that there was strong evidence that the mean number of insects that will be trapped differs between the different-colored traps. The outlier increased the sum of squares of error considerably, and this results in a much smaller value of *F*.
(c)

Group	Count	Mean	Std. dev.
Lemon yellow	6	114.667	164.416
White	6	15.6667	3.32666
Green	6	31.5000	9.91464
Blue	6	14.8333	5.34478

The unusually large values of the mean and standard deviation might indicate that there was an error in the data recorded for the lemon yellow trap.

14.81 (**a**) The pattern in the scatterplot is a roughly linear, decreasing trend. (**b**) The test in regression that tests the null hypothesis that the explanatory variable has no linear relationship with the response variable is the t test of whether or not the slope is 0. (**c**) Using software we obtain the following:

Variable	Coefficient	Std. error of coeff.	*t* ratio	*P*-value
Constant	4.36453	0.0401	109	≤ 0.0001
Number of promotions	−0.116475	0.0087	−13.3	≤ 0.0001

The t statistic for testing whether the slope is 0 is -133, and we see that P-value is ≤ 0.0001. Thus, there is strong evidence that the slope is not 0. The ANOVA in Exercise 14.49 showed that there is strong evidence that the mean expected price is different for the different numbers of promotions. this is consistent with our regression results here, because if the slope is different from 0, the mean expected price is changing as the number of promotions changes.

In this example the regression is more informative. It not only tells us that the means differ but also gives us information about how they differ.